INTERCONNECTION AND INSPECTION OF GRID CONNECTED ROOFTOP SOLAR PHOTOVOLTAIC SYSTEMS

DISCLAIMER

INTERCONNECTION AND INSPECTION OF GRID CONNECTED ROOFTOP SOLAR PHOTOVOLTAIC SYSTEMS

A Guide for DISCOM Engineers and Managers

First edition published 2018
by Routledge
4 Park Square, Milton Park, Abingdon, Oxon, OX14 4RN

and by Routledge
711 Third Avenue, New York, NY 10017

Routledge is an imprint of the Taylor & Francis Group, an informa business

British Library Cataloguing-in-Publication Data
A catalogue record for this book is available from the British Library

Library of Congress Cataloging-in-Publication Data

Names: Khanna, Ronnie, author.
Title: Interconnection and inspection of grid-connected rooftop solar photovoltaic systems : a guide for DISCOM engineers and managers/ principal authors, Ronnie Khanna, Ranjit Chandra, Arvind Karandikar, Deepanker Bishnoi.
Description: Abingdon, Oxon ; New York, NY : Routledge, 2018. | Includes bibliographical references and index.
Identifiers: LCCN 2018035222 | ISBN 9781138341289 (pbk.)
Subjects: LCSH: Building-integrated photovoltaic systems–Installation–Handbooks, manuals, etc. | Interconnected electric utility systems–India.
Classification: LCC TK1087 .K48 2018 | DDC 690/.83704724--dc23
LC record available at https://lccn.loc.gov/2018035222

ISBN: 9781138341289 (pbk)
ISBN: 9781003260509 (ebk)

Typeset in 11/13 Calibri
by Glyph Graphics Private Limited, Delhi – 110 096

DEDICATION

This handbook is dedicated to all the aspiring candidates and professionals who desire to achieve specific skills, which would be a lifelong asset for their future endeavours and create a niche for them in the solar PV sector.

CONTENTS

LIST OF FIGURES AND TABLES

FIGURES

TABLES

CONTRIBUTORS

This handbook is prepared jointly by the USAID-MNRE Partnership to Advance Clean Energy-Deployment Technical Assistance (PACE-D TA) Programme and the Skill Council for Green Jobs.

Principal Authors
Ronnie Khanna
Ranjit Chandra
Arvind Karandikar
Deepanker Bishnoi

Contributing Author & Chief-Editor
Tanmay Bishnoi

Skill Council for Green Jobs
Dr. Praveen Saxena

USAID Programme Manager
Anurag Mishra

CERTIFICATE

Skill India

SCGJ SKILL COUNCIL FOR GREEN JOBS

N·S·D·C
National
Skill Development
Corporation
Transforming the skill landscape

Certificate

COMPLIANCE TO
QUALIFICATION PACK – NATIONAL OCCUPATIONAL STANDARDS

is hereby issued by the

SKILL COUNCIL FOR GREEN JOBS

for

SKILLING CONTENT : PARTICIPANT HANDBOOK

Complying to National Occupational Standards of
Job Role/ Qualification Pack: '**Rooftop Solar Grid Engineer**' QP No. '**SGJ/Q0106 NSQF Level 5**'

Dr. Praveen Saxena

Authorised Signatory
(Skill Council for Green Jobs)

Date of Issuance: 14/05/2018
Valid up to*: 01/06/2019

*Valid up to the next review date of the Qualification Pack or the
'Valid up to' date mentioned above (whichever is earlier)

FOREWORD

India aims to deploy 175 gigawatts of renewable energy capacity by 2022, of which 100 gigawatts will be through solar energy. A substantial portion, 40 gigawatts, of this total will be through rooftop solar. Achieving this target will put India firmly on the path towards sustainable economic growth. This would also give India and its citizens the opportunity of creating and benefiting from new clean energy markets, bring down energy costs, generate livelihoods and create business opportunities while moving India towards an energy secure future.

Achievement of these targets depends heavily on how solar energy deployment is scaled. Human resource development, especially that of key stakeholders like utilities, financial institutions, developers, and entrepreneurs will be critical to the achievement of national targets and sustainable growth of the sector. The need for human resource development through training and capacity building becomes all the more important for stakeholders who are critical for ecosystem development while not directly being a part of the solar industry. This is especially true for engineers of distribution utilities who provide interconnection to newly installed rooftop solar systems, undertake site inspections and ensure the safety of the grid as well as the consumers. To meet the requirement of installing thousands of systems simultaneously, training and capacity of utility engineers needs to be built.

Skill Council for Green Jobs and the United States Agency for International Development (USAID), under the U.S.-India Partnership to Advance Clean Energy Deployment (PACE-D) TA Program, worked together to design a dedicated training program for utility engineers. This customized, standardized training program provides a comprehensive understanding of the rooftop solar sector, including technical and process related issues and practical insights.

Distribution utilities can benefit from this program and it is highly recommended to undertake an RPL certification from Skill Council for Green Jobs, aligned to the National Qualification of Rooftop Solar Grid Engineer (SGJ/Q0106) for implementing the skills gained during the program for interconnection and inspection of rooftop solar systems. More than ten distribution utilities have already taken the initiative and got junior and middle management officers trained under the qualification and are benefiting from the knowledge and skills acquired during the program.

This handbook on Rooftop Solar Grid Engineer is designed to be a resource for utility engineers working on and evaluating the development of grid-connected rooftop solar projects. It provides all required information for utility engineers to understand opportunities and challenges around the integration of solar onto the Grid. Information provided includes protocols and processes for connecting rooftop solar systems to the grid, compliance with applicable codes and standards, ensuring safety of the grid and its users, and overall policy and regulatory frameworks. It encompasses all relevant topics for utility engineers and provides the readers with all the necessary knowledge and skills. The contents of this book are in simple language, with the right amount of detail and calculations, while still providing the best understanding for inspection and interconnection of solar projects. We hope utility engineers contributing to the national rooftop solar target will benefit from this resource.

Anurag Mishra
Energy Team Leader
USAID India

Dr. Praveen Saxena
Chief Executive Officer
Skill Council for Green Jobs

PREFACE

Government of India is aiming to generate about 100,000 MW from solar energy by the year 2022. This includes a capacity of 40,000 MW to come from the rooftops of various buildings and houses spread throughout the country. To meet this ambitious goal and considering the huge requirement of integration skills for connecting the rooftop solar PV system with the grid, Skill Council for Green Jobs is targeting a special skilling course on Rooftop Solar Grid Engineer, which has been declared as a national qualification on approval of National Skills Qualification Committee, from Ministry of Skill Development and Entrepreneurship, Government of India.

This hand book on interconnection and inspection of grid-connected rooftop solar photovoltaic systems deals with the subject of how an individual can carry out pre-commissioning inspection of the grid-connected rooftop solar PV power plant and post-commissioning testing of the grid-connected rooftop solar PV power plant while maintaining personal health and safety at project site.

The major topics covered are as follows:

A. SGJ/N0118: Pre-Commissioning Inspection of the Grid-Connected Rooftop Solar PV Power Plant
    - Identify the key regulatory parameters for interconnection and metering arrangement including power quality of the grid at the project site
    - Identify and verify the documents required for connecting the rooftop solar PV power plant to the grid
    - Verification of the capacities, components, and safety protections of the rooftop solar PV power plant as per the relevant policies and regulations
B. SGJ/N0119: Post-Commissioning Testing of the Grid-Connected Rooftop Solar PV Power Plant
    - Verify the operation of the installed solar-metering system including import and export of energy
    - Test, record, and verify the power quality of rooftop PV power plant at the time of interconnection including harmonics, current, voltage, etc.

- Test and verify the inverter operation including anti-islanding functionality, overload, disconnect protections, power factor, and other relevant parameters

C. SGJ/N0106: Maintain Personal Health and Safety at Project Site

- Establish and follow safe work procedure
- Use and maintain personal protective equipment
- Identify and mitigate safety hazards
- Demonstrate safe and proper use of required tools and equipment
- Identify work- safety procedures and instructions for working at heights

The handbook has been envisaged to provide the readers with the knowledge and skills required on checks, audits, inspects, interconnects, and tests on different components of the grid-connected Solar PV Power Plant in compliance with all relevant codes, standards, and safety requirements and enable them to actively participate in the growing Indian solar market. The contents of this book are in simple language covering practical aspects, without going into too much theoretical details.

ACKNOWLEDGEMENT

This handbook on interconnection and inspection of grid connected rooftop solar photovoltaic systems is prepared by the Skill Council for Green Jobs and USAID Partnership to Advance Clean Energy-Deployment Technical Assistance (PACE-D TA) Programme to enhance the knowledge and skills of professionals from distribution companies and other professionals dealing with interconnection and inspection of grid- connected rooftop solar photovoltaic systems.

This handbook is aligned with the National Qualification on "SGJ/Q0106:Rooftop Solar Grid Engineer", and to make it wholesome and relevant, the national qualification had received inputs and feedback from the distribution companies, solar developers, EPCs, and other stakeholders, and these insights were used in detailing the tasks and skills required by an individual for connecting the rooftop solar photovoltaic system to the grid.

The USAID PACE-D TA Programme and Skill Council for Green Jobs would like to express their gratitude to all the individuals, organisations, and the key solar industry players who have provided valuable contributions in the preparation of this handbook. We would also like to thank the team of Routledge who guided the manuscript into production with great care and competence.

UNIT 1: ROLE OF ROOFTOPSOLAR GRID ENGINEERS

More and more states in India are launching rooftop solar PV programmes as part of policies and regulations related to net metering or net-metering or gross-metering framework. In rooftop solar programmes Discom engineers have major role to play as these systems get connected to Discom grid and interact on a daily basis. There are certain aspects of this grid connection programme that are important for these engineers.

- Most of these systems would be connected at Low Tension (LT) or Low Voltage (LV) grids
- The existing 'consumers' will turn into 'producers and consumers', sometimes known as 'prosumers'
- PV systems would generate power only during day time for about 10 to 12 years
- Solar PV generation is variable in nature and will change substantially during day time as well as over a year
- The distribution transformers may experience reverse current flow during PV generation time, if consumption at LV side is lower
- In case there are numerous small PV systems fitted on single phase, there may be improvement or increase in phase imbalance in part of the grid
- In case numerous PV systems are fitted on tail end of the grid, there is a possibility of voltage becoming firmed up, which otherwise drops as one goes away from the transformer. This may actually support voltage at the tail end
- The metering, billing, and accounting methods will change for those consumers who have installed rooftop systems

In the event of number of such systems on grid increasing day by day, the grid engineers may have to keep changing their roles to suit the modification. Some changes that are visible now are stated below:

- ➢ LT / LV side monitoring may be essential to understand generation and consumption patterns
- ➢ Communicating with inverters of some of the larger rooftop systems may be crucial to monitor (and, in future, control) generation from these systems

➢ Forecasting and scheduling for some of these systems may be necessary

➢ Capacity limits per distribution transformer may be increased from current levels

➢ Grid enhancement may be planned to accommodate more PV capacity

Discom is the implementation agency of grid connection rooftop solar PV programme, and on declaration of state rooftop solar PV regulation, either on net metering or gross metering, it should design and announce its own interconnection processes with forms and formats. This enables interested consumers to apply for rooftop solar PV system and also provides grid engineers the guidelines for interconnection and inspection of installed systems. A typical process and roles of different agencies are presented below. Grid engineers' involvement and responsibility would be at two stages, first at the time of approving consumer application and second at the time of synchronizing installed system with the grid. The work steps for these two stages are shown in different colours in Figure 1.1.

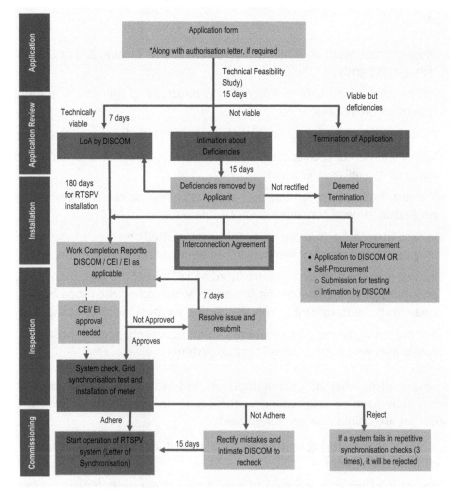

Figure 1.1: *A typical interconnection process of Rooftop Solar PV System*

Once the consumer applies for net metering (or gross metering, if available), first step for the engineer is to check technical feasibility and approve or reject the application based on facts in the field. At this stage, the engineer checks the regulatory limits in the particular state and confirms that the application is within these limits. These conditions are:

- Applied capacity with respect to the consumer's existing connection in terms of contract demand (CD) or sanctioned load (SL) – the condition can be the consumer proposed PV plant capacity capped at certain percentage of CD/SL, like 100%, 90% or 80%, and so on
- Cumulative capacity limit of all applications under a single distribution transformer (DT) can be capped at 80%, 75%, 40%, 30%, 15%. There would be up-to-date register maintained at the local office of the Discom (sub-division and / or division office) of all approved rooftop PV applications under each DT so that the newly received application can be easily checked for this cumulative capacity limit

Additionally, Discom may get clearance from its finance department for the applicant/consumer not being a defaulter. In such a case, the consumer is normally given a chance to clear the default and resubmit the application.

The field engineer may conduct a site visit before approving or rejecting the application and interact with the applicant/consumer and/or installer to understand feasibility of plant on site and also to check present interconnection, metering, and DT to which the consumer is connected.

This visit is a preliminary site inspection and may be avoided in case records are well maintained at the local-level office.

Once the consumer application is approved by the Discom through a feasibility approval letter, the consumer shall install the system under guidelines provided in the approval letter as well as in the state regulation and Discom interconnection process. There needs to be two types of inspection by the Discom (grid) engineer at this stage once work completion report is received from the consumer. All states provide for compulsory inspection and approval by state Electricity Inspectorate (EI) for certain capacity of rooftop PV plant. In cases where such prior approval is mandatory, the Discom engineer must carry out its own inspection only after the EI has approved the particular installation. This can be done either by Chief Electrical Inspector or Electrical Inspector or any other officer within the EI depending on the state guidelines for different capacities of the plant.

Grid engineers' role and responsibilities in these steps are detailed in this guide in subsequent sessions.

UNIT 2: INTRODUCTION TO ROOFTOP SOLAR PV SECTOR IN INDIA

2.1 AN OVERVIEW OF RENEWABLE ENERGY AND SOLAR SECTOR IN INDIA

2.1.1 Renewable Energy Technologies

'Renewable Energy' refers to the grid quality electricity generated from renewable energy sources. 'Renewable Energy Power Plants' refers to the power plants other than the conventional power plants generating grid quality electricity from renewable energy sources. 'Renewable Energy Sources' means renewable sources such as solar, wind, small hydro, biomass, including its integration with combined cycle, biomass,bio fuel cogeneration, urban or municipal waste, and other such sources as approved by the MNRE.

Renewable Energy is derived from the natural processes that are replenished constantly. In its various forms, it derives directly from the sun, or from heat generated deep within the earth. Included in the definition is electricity and heat generated from solar, wind, ocean, hydropower, biomass, geothermal resources and biofuels and hydrogen derived from renewable resources. (Source: International Energy Agency)

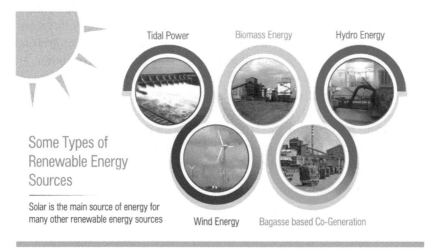

Figure 2.1: Renewable energy technologies (Tata BP Solar Ltd.)

In the following sections, we discuss about the different renewable energy sources that can provide us energy for sustaining our civilisation:

Solar Energy

Solar energy which is being received as heat and light can be converted into thermal energy and electricity. 'Solar PV power' means the solar photo voltaic power project that uses sunlight for direct conversion into electricity through photo voltaic technology. 'Solar thermal power' refers to the solar thermal power project that uses sunlight for direct conversion into electricity through concentrated solar power technology.

Biomass Energy

Biomass energy is a term that includes all energy materials derived from biological sources, including wood wastes, agricultural residues, food industry wastes, sewage, municipal solid waste (MSW), and dedicated herbaceous or woody energy crops. 'Biomass gasification' is a process of incomplete combustion of biomass resulting in production of combustible gases consisting of a mixture of Carbon monoxide (CO), Hydrogen (H2), and traces of Methane (CH4), which is called producer gas.

Wind Energy

Energy harnessed from the velocity of wind is known as wind energy. Wind technologies convert the energy of moving air masses at the earth's surface to rotating shaft power that can be directly used for mechanical energy needs (e.g., milling or water pumping) or converted to electric power in a generator. Two major types of turbines exist and they are defined based on the axis of blade rotation: horizontal-axis (which currently dominate commercial markets) and vertical-axis turbines. Evaluating the quality of wind resource at a specific location is critical in determining the suitability of wind turbines.

Hydro Energy

Hydropower facilities exploit the kinetic energy in flowing or falling water to generate electricity. Conventional hydropower facilities use water from a river, stream, canal, or reservoir to continually produce electrical energy, and water releases from single-purpose reservoirs (i.e., dedicated to power production) can be quickly adjusted to match electricity loads.

Geothermal Energy

The energy harnessed from the stored thermal energy below the earth's surface is known as geothermal energy. We can also use geothermal energy to make electricity. A geothermal power plant works by harnessing steam or hot water reservoirs underground; the heat is used to drive a generator to produce electricity.

Wave Energy

Wave energy is produced when electricity generators are placed on the surface of the ocean. The energy provided is most often used in desalination plants, power plants, and water pumps. Energy output is determined by wave height, wave speed, wavelength, and water density.

A number of technologies have been developed as alternative energy technologies that supplement our energy needs through renewable and environmentally friendly energy sources on the background of fast-depleting fossil fuels and their adverse impact on environment.

Solar Radiation

The energy produced by the sun is emitted to the PV system primarily as radiation. The radiation that falls directly on the earth's surface is termed as direct radiation. The fraction of solar radiation that is reflected back into space is known as the albedo of the atmosphere. And the radiation that passes through clouds and gases in the atmosphere is known as diffused radiation.

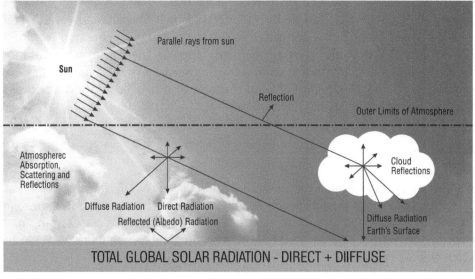

Figure 2.2: Types of solar radiation (Tata BP Solar India Ltd)

Movement of Sun Across the Sky

Solar radiation available at a particular location keeps changing both on a daily basis as well as annually. Hence, the amount of solar radiation keeps fluctuating at any given time interval depending on the position of the sun and the location of the site. The three main cases of solar radiation depending on the position of the sun can be categorised with the help of the solstices. The winter solstice occurs on 21st December. The Equinox is when the sun is directly over the equator and occurs twice in a year on 21st March and 21st September and the summer solstice occurs on 21st June.

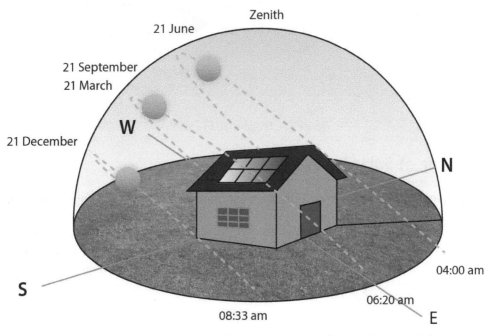

Figure 2.3: Orientation of the sun throughout the year (www.solarpraxis.de)

Working of solar thermal system:

The basic principle of solar thermal heating is to utilise the sun's energy and convert it into heat. Solar thermal heating systems harvest solar energy through a collector which absorbs energy from the sun to heat a fluid. Direct systems circulate water through solar collectors where it is heated by the sun. The heated water is then stored in a tank or used directly. These systems are preferable in climates where it rarely freezes. For low temperature applications, solar flat plate collectors (FPC) or evacuated tube collectors (ETC) are used.

Evacuated Tube Diagram

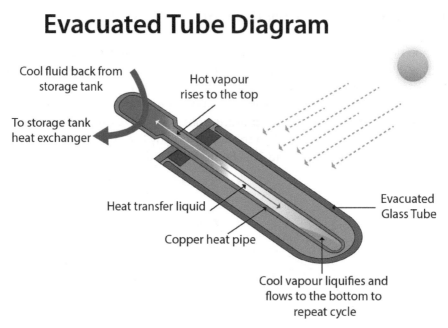

Figure 2.4: Diagram of ETC system

Working of Solar PV System

Solar panel works by allowing photons, or particles of light, to knock electrons free from atoms, generating a flow of electricity. Solar panels actually comprise many smaller units called PV cells. These cells are made of two different types of semiconductors that are joined at p-n junction. These p-n junctions consist of an energy barrier that requires a minimum threshold of photons to exceed the barrier potential in order to generate electricity.

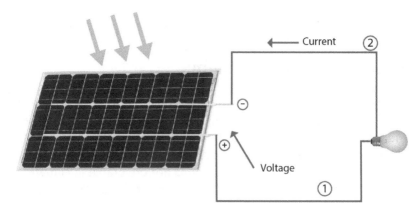

Figure 2.5: Operation of a PV module

2.1.2 Solar Energy Technologies

As discussed above, the solar energy technologies can be classified into solar thermal energy technologies and solar PV energy technologies. Based on the above two technologies, several solar energy systems and devices have been developed over the years. The systems, which are now available commercially, are described below:

● Off-grid systems

As the name implies, these systems are not connected to electricity grid. So these systems can ideally be used for lights, fans, mobile telephone charging, water pumping, water purification refrigeration, etc. in remote isolated areas and also in areas where grid availability is significantly low.

Figure 2.6: Off-grid Solar PV applications

● Small distributed applications

A solar PV system is classified into small distributed applications if the system size falls under 20kWp. These solar power units can be installed on ground or on the rooftop of a building to meet the electricity requirement of the community or just the building. These can operate in off-grid mode where the electricity produced during the day (when sun shine is available) can be stored in battery and can be used as and when required. These systems can also be grid connected. In grid interactive mode, the electricity produced is first utilised for the local load and the excess can be transported to the grid. Similarly, when the electricity generated onsite is not sufficient or is not available, then electricity from the grid is used for powering the loads. Such systems are now very common as rooftop solar systems.

Figure 2.7: Small distributed Solar PV applications

● **Large distributed applications**

These are typically installed on buildings with large roof areas, such as warehouses, industrial or commercial buildings. The systems that fall under this category are generally from a few kilowatts to 1MWp. These systems can also operate in off-grid mode with battery storage or in grid interactive mode. Grid-based systems are now vigorously promoted in the country.

Figure 2.8: Large distributed Solar PV applications

- ## Ground-mounted utility scale grid-connected power plants

Utility scale projects tend to be of 1MWp or above and are generally ground mounted since they occupy a very large area. These large PV systems tend to be directly connected to the grid.

Figure 2.9: Renewable energy technologies

● Microgrids

Micro grids essentially consist of a number of energy generation sources that cater to interconnected loads and can be operated in parallel or independent of the utility grid.

For example, numerous houses can be connected to a number of solar PV modules For example, a project where number of houses are connected to solar PV modules operating independently, and charging battery banks can be classified as a micro grid.

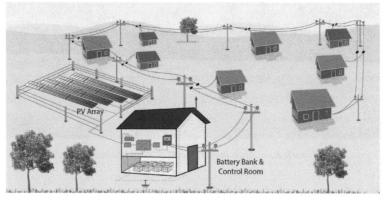

Figure 2.10: *Micro grid Solar PV applications (www.solarpraxis.de)*

● Building integrated PV systems

As the name suggests, solar PV modules are integrated into the building architecture in an aesthetic manner for onsite generation of electricity.

Figure 2.11: *Building Integrated PV Systems*

Solar Water Heating Systems

Solar water heating systems can heat water by using solar energy. There are many types of solar water heaters. The most common systems are FPCs and ETC systems. The FPCs are metallic collectors with a glass cover over the selectively coated absorbers. The absorbers are mainly of tube and fin type. In case of ETC systems, the water piping is surrounded by two concentric tubes of glass with a vacuum in between to provide necessary heat insulation. In both the cases, the cold water enters through the top of the structure, passes through the tubes, and gets heated up in the process. The heated water, as a result of thermosiphon effect, is transported to the storage tank.

Figure 2.12: Solar Hot Water Application

Concentrated Solar Power (CSP)

Concentrating solar power technologies use different mirror (reflector) configurations to concentrate the available solar radiation onto a receiver and convert it into thermal energy at high temperature. The thermal energy can then be used for generation of steam to drive a turbine to produce electrical power or used as industrial process heat.

Figure 2.13: Concentrated Solar Power applications

Passive Solar Design

Passive solar design integrates the architecture of the building with the movement of the sun across the sky in order to provide natural heating from the sun in winter and shading in summer months.

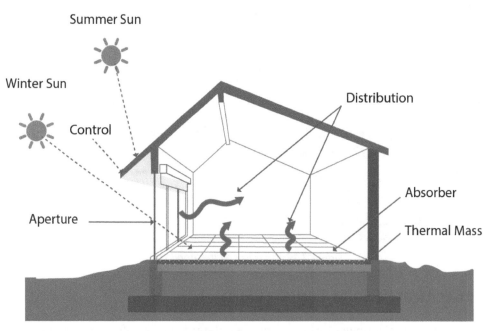

Figure 2.14: Five elements of passive solar design

2.2 AN OVERVIEW OF ROOFTOP SOLAR SECTOR IN INDIA

2.2.1 Growth of Solar Power Installation in India

The impact of fossil fuel-based energy system in creating and exacerbating climate change has been widely established and accepted. Governments and communities across the world, including India, are investing in renewable energy technologies to replace existing fossil fuel-based energy generation systems with cleaner sources.

India is a relatively new entrant in the renewable energy industry, and it is reflected in its energy generation mix, where fossil fuel-based sources account for almost 58% of total sources of generation. Upto June 2017, renewable energy, excluding large hydropower, was approximately 17.4% of the total energy mix, i.e., installed capacity of 57.472 GW out of total installed capacity of 329.4 GW. Of all the forms of renewable energy in India, solar power has received a great level of interest from public and private sectors as one of the solutions to India's energy challenges. With falling prices and increasing demand of solar energy, renewable energy mix has gone upto almost 18% out of total installed capacity of 340.53 GW (as on 12 April 2018) as given in the Figure 2.15.

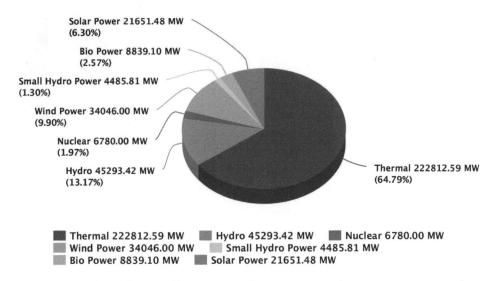

Figure 2.15: *Share of Renewable Energy in Indian Power Sector (Source:* www.npp.gov.in*)*

Use of solar power as a supplementary source to traditional grid-based electricity is relatively recent in India. The Generation Based Incentive (GBI) scheme, announced in January 2008, was the first step by the government to promote installation of grid- connected solar power plants. Prior to this, the Rural Electrification Programme of 2006 is regarded as the first set of guidelines set by the Indian Government. This programme was focused on promoting off-grid solar applications including solar lanterns, solar pumps, home lighting systems, street lighting systems, and solar home systems.

In 2008, the Indian Government announced National Action Plan for Climate Change (NAPCC), which laid the basis of National Solar Mission (NSM) launched in January 2010.

The NSM set targets, provide policy guidelines, and financial incentives to support the growth of solar power in India. Under NSM, the installed capacity of solar power in India has propelled from just over 2 MW in 2009 to 20,006 MW as on 30th Jan2018, as illustrated in Figure 2.16.

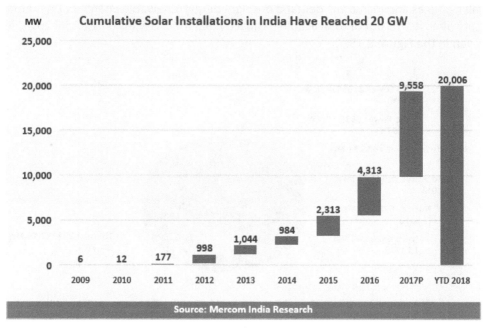

Figure 2.16: Installed solar capacity (MW) in India

2.2.2 India's National Solar Mission

The NSM is an initiative of the Government of India and the state governments. It commenced in 2010 to address India's energy security challenges while promoting ecologically sustainable growth. The NSM will be the major contribution to India's intended commitment to meet its share of climate change mitigation and adaptation activities agreed within the United Nations Framework Convention on Climate Change (UNFCCC).

The UNFCCC is an international treaty, established to coordinate international efforts in addressing the challenges of climate change.

The overall objective of the NSM is to make solar cost completive with fossil fuel-based energy options by 2022 through the following actions:

- Long-term policy
- Large-scale deployment goals
- Aggressive research and development
- Domestic production of critical raw materials, components, and products

Following up on the objectives of the mission, the Government of India announced ambitious targets for grid-connected solar power. Under these targets, a total of 100,000 MW of grid-connected solar power is to be installed across India by the year 2022. Utility scale solar and rooftop solar will contribute 40% each towards this target, while entrepreneurs will get an opportunity to meet 20% of the targets, with a focus on local job creation and skill development. A detailed breakdown of this target is illustrated in Figure 2.17:

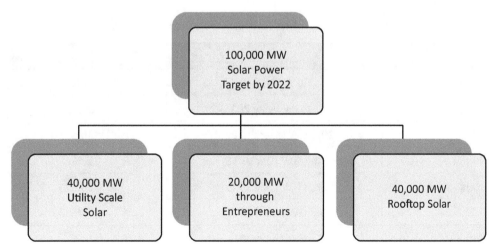

Figure 2.17: Targets under India's National Solar Mission

States have been allocated their contribution towards the national targets by Ministry of New and Renewable Energy. As illustrated in Figure 2.18, Indian states have been allocated overall targets of 40,000MW rooftop solar separately to be completed by 2022.

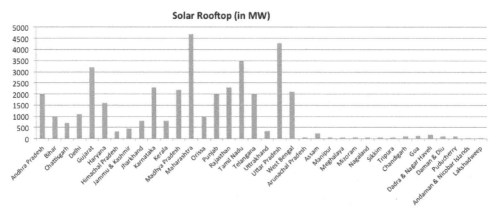

Figure 2.18: *Solar targets allocated to states*

2.2.3 Global Scenario of Solar Power Projects and Share of Rooftop PV

Rooftop solar power projects are very popular across many countries. Germany has the highest rooftop solar installed capacity (~24GW), while Australia has the highest percentage of rooftop solar installations (80%) as the share of total solar power installations including utility scale. A list of countries leading in rooftop installations is provided in Table 2.1.

Table 2.1: List of countries leading in rooftop solar installations

Country	Total PV Installed Capacity	Share of Rooftop PV (%)
China	50.3 GW (April 2016)	~ 20%
Germany	40.1 GW (May 2016)	~60%
US	29.3 GW (March 2016)	~ 40%
Japan	34.4 GW (Dec 2015)	~35%
Italy	18.9 GW (Dec 2015)	~ 70%
UK	9.8 GW (April 2016)	~ 52%
India	7.8 GW (June 2016)	~ 10%
Spain	5.4 GW (Dec 2015)	~ 70%
Australia	5.1 GW (Dec 2015)	~ 80%

Source: REN21 Report

2.2.4 Rooftop Solar PV Installation Status

The total installed capacity of grid-connected rooftop solar power in India at the time of writing this report is 1,861 MW across different sectors as reported by Bridge to India in their India Solar Rooftop Map 2017, published in September. Out of which, majority of installations have happened in industrial sector with around 798MW of installed capacity followed by 393MW in commercial establishments, 377 in residential sector, and 294MW in public sector entities. State-wise achievements can be found at http://www.bridgetoindia.com/wp-content/uploads/2017/12/India-Solar-rooftop-Map-Dec.pdf

2.3 FEATURES OF ROOFTOP SOLAR PV SYSTEMS

Grid-connected solar power systems use sunlight as fuel to generate DC power. This DC power is converted to AC power via the inverter, which is then fed into the electricity grid. This is the fundamental working principle of all grid-connected systems.

A large solar PV array is directly connected to the transmission and distribution system, which then carries the generated electricity through the distribution network to the end user. The size of a utility scale system can be as small as 50kWp, while systems as large as 250MWp have been installed in recent years in the United States. In India, the State of Gujarat is host to world's second largest PV plant with nominal capacity of 214MWp. Under the NSM, India is going to have ultra MW scale PV power parks in the coming years.

In rooftop solar power projects, the solar array is placed on the shadow-free and stable south facing (north facing in southern hemisphere) roof area of a residential, commercial, or industrial establishment. The power generated by a rooftop PV system is typically consumed by the loads within the building, and excess power is then fed into the grid. The quantity of exported power is subject to policies within the respective states and metering restrictions.

The size of a rooftop PV project can be as small as 1kWp to 5kWp for a residential roof and as large as 100kWp to 500kWp and even MW capacity, subject to roof availability. At the time of writing, the largest rooftop solar scale project commissioned in India and the world is 12MWp commissioned in Amritsar (Punjab).

While both utility and rooftop PV projects serve at different generation scales and point of power delivery, they broadly achieve similar goals in the context of the overall energy supply system. The key advantages of using rooftop projects, where possible, over utility projects is that it removes the typical transmissions and distribution losses in the system and also optimises the use of available infrastructure. A summary of key advantages of rooftop PV over utility scale solar power projects is illustrated in Table 2.

Table 2.2: Benefits of rooftop solar power projects over utility scale projects

Solar Deployment Areas/Benefits	Land Requirement	T&D Loss	Investments	Optimal Utilisation of Infrastructure
Large ground-mounted solar projects	Dedicated	Injected to national grid	Need large investments	Need new transmission lines & enhanced grid capacity
Large solar rooftop projects (industrial/ commercial/ institutional)	Un-utilised roofs	Consumed at generation point	Can mobilise medium investments	Infrastructure expansion needs can be minimised
Small rooftops projects (residential)			Can mobilise small investments	

UNIT 3: ROOFTOP SOLAR PV ECOSYSTEM

3.1 AN OVERVIEW OF ROOFTOP SOLAR PV TECHNOLOGY

3.1.1 Terminologies Used in Solar Industry

Radiation – The electromagnetic energy which is produced by the sun and transmitted through photons is termed as solar radiation.

Solar Irradiation (Energy) / Insolation – It is the sun's radiant energy incident on a surface of unit area. Irradiance is measured in Watt-hour/m^2 or kW-h/ m^2.

Equinox – Literally 'equal night', a day when the number of hours of daylight equals the number of hours of night.

The vernal equinox, usually March 21, signals the onset of spring, while the autumnal equinox, usually September 21, signals the onset of autumn.

Solstice – A day when the sun is at the highest point in the sky (summer solstice, 21 June) or at the lowest point in the sky (winter solstice, 22 December).

Zenith Angle – The angle between the direction of interest (of the sun, for example) and the zenith (directly overhead).

Azimuth Angle – The angle between the horizontal direction (of the sun, for example) and a reference direction (usually north, although some solar scientists measure the solar azimuth angle from due south).

Elevation – The elevation angle is the angular height of the sun in the sky measured from the horizontal. The elevation is 0° at sunrise and 90° when the sun is directly overhead.

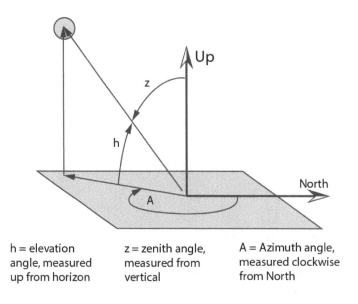

h = elevation angle, measured up from horizon

z = zenith angle, measured from vertical

A = Azimuth angle, measured clockwise from North

Figure 3.1: Solar clockwise from north

Angle of Incidence – The angle that a ray (of solar energy, for example) makes with a line perpendicular to the surface. For example, a surface that directly faces the sun has a solar angle of incidence of zero, but if the surface is parallel to the sun (for example, sunrise striking a horizontal rooftop), the angle of incidence is 90°.

Air Mass –Air mass is the path length which light takes through the atmosphere normalised to the shortest possible path length (i.e., when the sun is directly overhead). The air mass measures reduction in the power of light as it passes through the atmosphere and is absorbed by air and dust. The air mass is defined as

$$\text{Air Mass} = 1/\cos(\text{theta})$$
Where Theta is the angle from vertical (Zenith Angle)

Solar Noon – The time at which the position of the sun is at its highest elevation in the sky. At this time, the sun is either due south (typically in the northern hemisphere) or due north (typically in the southern hemisphere).

Solar Thermal Technology: It refers to the direct conversion of sunlight energy in heat energy.

Solar Photovoltaic (PV) Technology: It refers to the direct conversion of sunlight energy into electrical energy.

Solar PV Cell: It is defined as the semiconductor device that directly converts sunlight energy into DC (Direct Current) electricity.

Solar PV Module: It is defined as the series connected assembly of solar PV cells to generate DC electricity.

Solar PV Array: It is defined as the connected (series/ parallel or both) assembly of solar PV modules to generate DC electricity.

Electrical Grid/Grid: It is defined as the interconnected network which delivers electricity from suppliers to consumers.

3.1.2 Types of Rooftop Solar PV Systems and Working Principles

As the name suggests, a rooftop solar PV system is located on the roof of a home, factory, shed, or any other building. The installation is secured with help of frames and structures. The technology works in the same manner as described in the earlier chapter – solar array converts photon particles in sunrays to DC current, which is changed over to AC current using inverter. Subject to the type of requirement/ consumption profile, a rooftop solar PV system may be classified into following three categories:

    I.   Grid-Connected Rooftop PV System

    II. Off-Grid Rooftop PV System

    III. Hybrid Grid-Connected Rooftop PV System

Sizing of the PV array depends on the following variables: load demand, demand profile, availability of structurally sound shadow-free south facing roof area, and availability of grid interconnection facilities.

I. Grid-Connected Rooftop Solar PV System

A grid-connected rooftop PV system interacts with the grid and there is no internal energy storage. Based on the metering arrangement, the electricity generated by the PV array is consumed by the loads first, and the excess power generated is exported to the grid. A typical representation can be seen in Figure 3.2.

Figure 3.2: Typical grid connected rooftop PV system (www.indiamart.com)

II. Off-Grid Rooftop Solar PV System

As the name suggests, an off grid rooftop PV system has no interaction with the grid. It is installed where there is absence of grid and hence it requires batteries for internal storage of energy for usage. A typical representation can be seen in Figure 3.3.

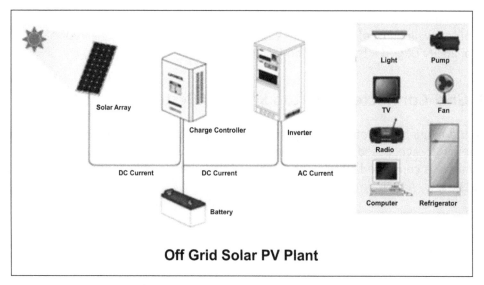

Figure 3.3: Off-grid rooftop PV system

III. Hybrid Grid-Connected Rooftop Solar PV System

This type of system is a combination of the first two system, i.e., it interacts with the grid and also has batteries for storage of energy for use when the grid or solar is not available. A typical representation can be seen in Figure 3.4.

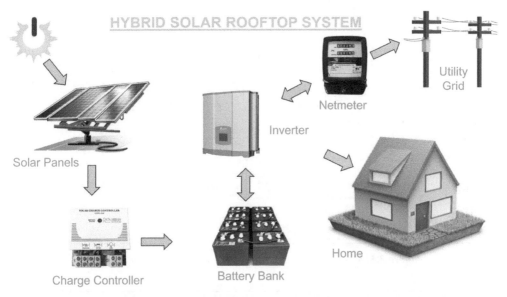

Figure 3.4: Typical hybrid grid connected rooftop PV system (www.go-solar.co)

3.2 DIFFERENT CONFIGURATIONS OF GRID-CONNECTED ROOFTOP PV SYSTEMS

There can be different types of rooftop PV systems even under grid connected framework. Following simple single line diagrams (Figures 3.5 and 3.6) present these different types.

In Figure 3.4, it should be noted that the consumer can be HT or LT consumer, but interconnection voltage of rooftop PV system under gross metering will depend on the capacity of the system and the interconnection point for such systems that need to be connected to HT voltages, due to its capacity, it can be outside but near to the consumer premises or can be far away from the premises.

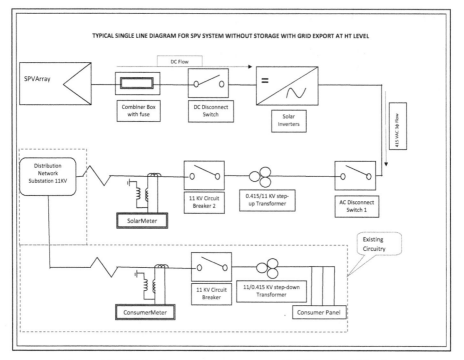

Figure 3.5: *SLD for Gross-metered system with HT-level grid feeding*

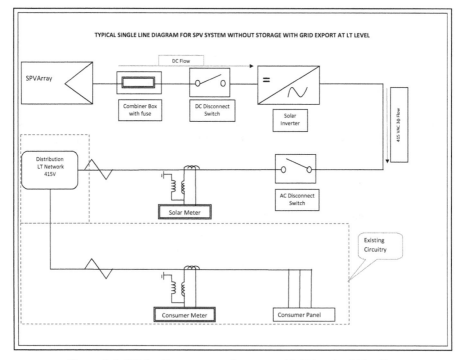

Figure 3.6: *SLD for Gross-metered system with LT-level grid feeding*

In Figure 3.6, it should be noted that the consumer can be HT or LT consumer, but interconnection voltage of rooftop PV system under gross metering will depend on the capacity of the system and the interconnection point at LT voltages, for such systems can be outside but near to the consumer premises or can be far away from the premises.

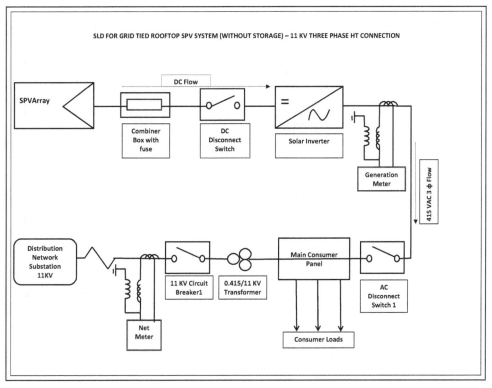

Figure 3.7: *SLD for net-metered system for existing HT consumer*

In Figure 3.7, it should be noted that the consumer can be HT or LT consumer, but interconnection voltage of rooftop PV system under net metering will always be at LT voltage level, either at single phase 230V or three phase 415V depending on the capacity of the system. The interconnection point for such systems shall always be inside the consumer premises at consumer's consumption voltage level. Only in case of HT consumer as above, the metering will still be at existing HT level.

In Figure 3.8, it should be noted that the consumer can be HT or LT consumer, but interconnection voltage of rooftop PV system under net metering will always be at LT voltage level, either at single phase 230V or three phase 415V depending on the capacity of the system. The interconnection point for such systems shall always be inside the consumer premises at consumer's consumption voltage level. In case of LT consumer as above, the metering also will be at existing LT level.

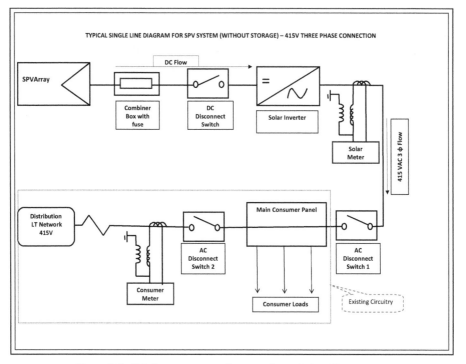

Figure 3.8: *SLD for net-metered system for existing LT consumer*

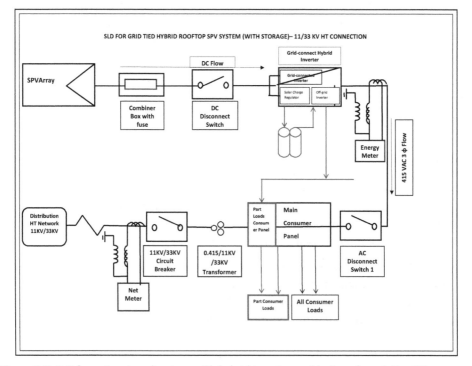

Figure 3.9: *SLD for net-metered system with hybrid inverter and battery for existing HT consumer*

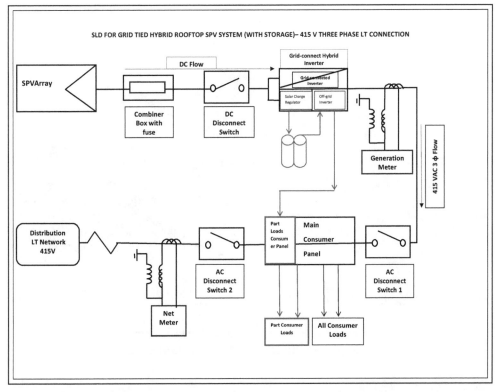

Figure 3.10: SLD for net-metered system with hybrid inverter and battery for existing LT consumer

3.3 COMPONENTS OF A SOLAR PV SYSTEM

The major components of a grid-connected rooftop solar PV system:

I.   PV modules

II.   Grid-connected inverter

III. Array mounting structure

IV. Balance of system components, including DC cables, AC cables, junction box, cables, fuses, isolators, safety components, such as earthing and lightning protection systems and batteries

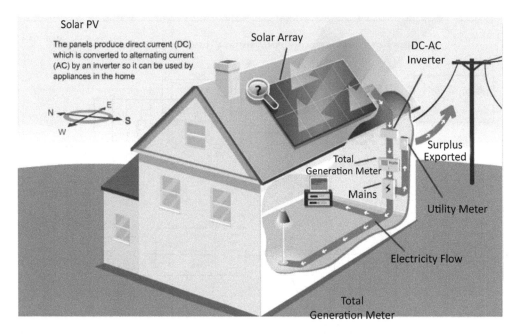

Figure 3.11: *Grid-connected rooftop PV system*
(http://atlaspowersolution.com/on-grid-pv-systems/)

Figure 3.12: *Typical rooftop Installations*

I. PV Modules

The solar PV is evolving and different technologies have been emerged in recent years. There are two major types of solar modules: crystalline and thin film modules.

Crystalline modules are classified into mono-crystalline and poly-crystalline based on the growth of silicon crystal. Thin film modules are classified based on compounds used in making the modules.

Some of the commonly used PV modules which are available in the market are summarised in Table 3.1.

Table 3.1: Different types of PV technologies key parameters

Type	Technology	Efficiency	Area (ft2/kWp)
Crystalline			
Mono-crystalline	• Ingot pulled from molten poly-silicon from furnace • Wafers cut from ingots	16–22%	65–49
Polycrystalline modules	• Brick of molten poly-silicon from furnace • Wafers cut from brick	16–19%	65–56
Thin film			
CdTe	Gaseous deposition on substrate (glass)	9–16%	120–65
CIGS	Gaseous deposition on substrate (glass)	9–15%	120–72

Figure 3.13: Different types of PV technologies

In selecting modules for an application, attention must be paid to its testing and compliance certificates evidencing meeting of relevant codes and standards. The particular information can be found from the RFID on the backsheet of solar modules, the details of which should match with the datasheet as submitted in the application as shown in Figure 3.14.

Figure 3.14: RFID Tag of Module and its Datasheet

Performance Degradation Over Life Cycle

The performance of a PV module will decrease over time. The degradation rate is typically higher in the first year upon initial exposure to light and eventually stabilises. Factors affecting the degree of degradation include the quality of materials used in manufacture, the manufacturing process, the quality of assembly, and packaging of the cells into the module, as well as maintenance levels employed at the site. Generally, degradation of a good quality module is considered to be about 20% during the module life of 25 years @ 0.7% to 1% per year.

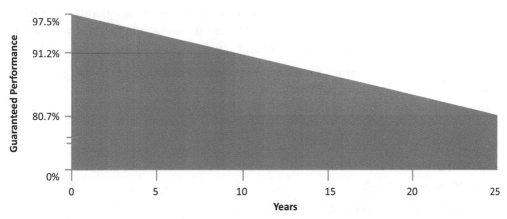

Figure 3.15: Example of PV module degradation

PV Module Performance and Qualification Standards Adopted by MNRE

- STC performance IEC60904-1
- IEC 61215 (Revised) Crystalline Silicon Modules
- IEC 61853 Energy Rating
- IEEE 1262 (Crystalline & Thin Film Modules)
- JIS C 8917 (Crystalline Silicon Modules)
- IS 14286: 1995 (Crystalline Silicon Modules)
- IEC 61646 (Thin Film Modules)
- 62108 Concentrator Photovoltaic (PV) receivers & modules design, qualification, and type approval
- 61730-I & II Module Safety Construction & Qualification

Performance of modules is tested under controlled conditions for uniform comparison of available technologies and different manufacturers. These conditions are call standard test conditions (STC), and based on this testing, modules are rated. STC conditions are as follows:

- Cell temperature 25˚C
- Irradiance of 1000W/m^2
- Air mass of 1.5

This implies, because measuring performance of a module cell temperature is taken into consideration which is different from ambient temperature. Cell temperature is always higher than ambient temperature and can go to 25°C higher ($T_{cell} = T_{air} + 25°C$).

The real-field conditions are different from STCs and hence most of the manufactures also test their equipments performance closer to real-world conditions known as Normal Operating Cell Temperature (NOCT) with the following conditions:

- Ambient air temperature 20˚C
- Irradiance of 800W/m2
- Wind speed 1m/s
- Electrically open circuit

II. Grid-Connected Inverter

Inverter changes DC electricity produced from the arrays into AC sine wave that matches the AC supply in voltage and frequency to which it is connected. The resulting power can then be used by the AC loads and also exported to the grid.

In a grid-connected PV system, the PV array is directly connected to the grid-connected inverter. These inverters cannot produce a grid equivalent AC sine wave independently, as the inverter must detect and reference the grid to be able to operate. If the AC grid is not present, the inverter will simply not function.

Grid-converted inverters can be classified based on the following criteria:

- Isolated or non- isolated
- Interfacing set-up in respect to PV array

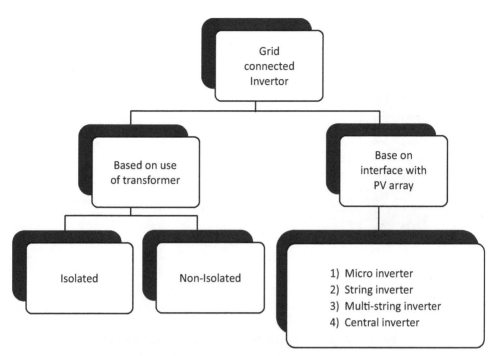

Figure 3.16: *Classification of grid-connected inverter*

Isolated Inverter: According to IEC 62548 standard, an Isolated inverter is an 'inverter where there is at least a simple separation between input and output circuits (e.g. by means of a transformer with separate windings)'.

Non-Isolated Inverter: According to IEC 62548 standard, a non-isolated inverter is an 'Inverter that does not have at least a simple separation between input and output circuits'. These inverters are known as transformer-less inverters.

Inverters that are classified based on their interfacing features are listed below.

Micro inverters or modular inverters are small isolated inverters (some can also have an isolating transformer) that are designed to be mounted on the back of the solar module.

String inverters range from 1kWp up to approximately 20kWp and are connected to a string of PV array. A 'string' is made of PV modules connected in series. An example of a string inverter is shown in Figure 3.18. Multiple inverters can be used to spread the generated power over more than one phase. Multiple inverters may also be used for extra redundancy or when a larger inverter is either not available or more expensive than two smaller inverters.

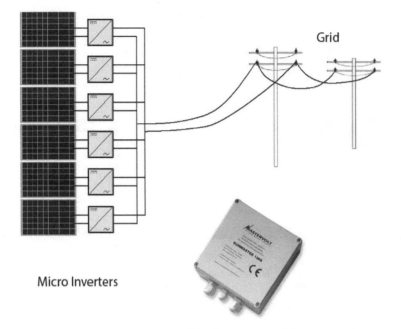

Figure 3.17: *Micro Inverters (also known as Modular Inverters)*

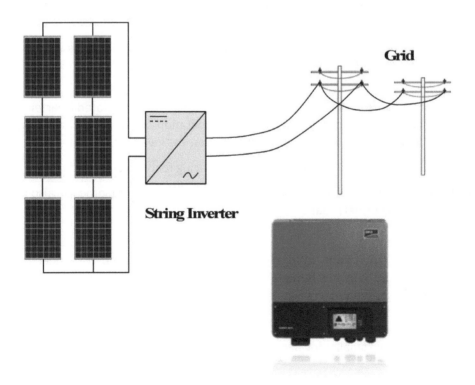

Figure 3.18: *Grid-connected string inverter*

Multi-sting inverter has a number of maximum power point tracker (MPPT inputs). This feature allows a solar array to be divided into multiple strings and connect each string individually to the inverter as shown in Figure 3.19.

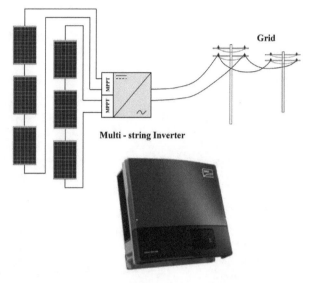

Figure 3.19: *Grid-connected multi-string inverter*

Central Inverters are used for large grid-connected solar systems. A central inverter is similar to a string inverter with multiple strings – the difference is that central inverters could have PV arrays divided into a number of sub-arrays each comprising of a number of strings. Central inverters are decided based on their operating voltage window specific to the inverter. An example diagram of a central inverter is shown in Figure 3.20.

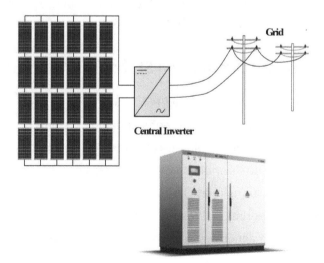

Figure 3.20: *Grid-connected central inverter*

Multi-Mode Inverters also called a hybrid inverters and they are used used with a gird-connected PV system with storage battery bank. This type of inverter works as a standalone inverter when there is no grid and grid, connected inverter stops feeding to the grid. They are generally more expensive than a normal grid-connected inverter.

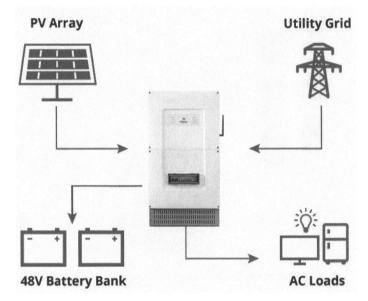

Figure 3.21: *Grid connected multi-mode inverter and storage battery*
*(*https://www.solarchoice.net.au/blog/redback-technology-home-battery-storage-solutions*)*

The types of batteries vary from manufacturer to manufacturer, and they all have their unique operating principles. Currently, lead acid batteries are most affordable and common for solar storage applications. Developments in advanced battery chemistries such as the lithium ion and ferrous derivatives will be providing affordable and more reliable storage solutions in the near future.

Solar PV Inverters – Quality and Standards Adopted by MNRE

- IEC 62109 – 1,2: Protection degree (IP65, IP54)
- IEC 61683: For measuring efficiency of standalone systems
- IEC 62891: Overall efficiency of grid-connected PV inverters
- IEC 62116: to measure Islanding prevention measures
- IEC 60255-27: Measuring relays and protection equipment
- IEC 60088 – 2: Environmental testing of inverters
- IEC 61000: Electromagnetic interference and electromagnetic compatibility testing

A sample name plate of inverter can be found on the inverter itself that can be matched with datasheet of the same model as shown in Figure 3.22.

Technical data	Sunny Tripower 10000TL	Sunny Tripower 12000TL
Input (DC)		
Max. DC power (@ cos φ = 1)	10200 W	12250 W
Max. input voltage	1000 V	1000 V
MPP voltage range / rated input voltage	320 V – 800 V / 600 V	380 V – 800 V / 600 V
Min. input voltage / initial input voltage	150 V / 188 V	150 V / 188 V
Max. input current input A / input B	22 A / 11 A	22 A / 11 A
Max. input current per string input A² / input B²	33 A / 12.5 A	33 A / 12.5 A
Number of independent MPP inputs / strings per MPP input	2 / A:4, B:1	2 / A:4, B:1
Output (AC)		
Rated power (@ 230 V, 50 Hz)	10000 W	12000 W
Max. apparent AC power	10000 VA	12000 VA
Nominal AC voltage	3 / N / PE; 220 / 380 V, 3 / N / PE; 230 / 400 V, 3 / N / PE; 240 / 415 V	3 / N / PE; 220 / 380 V, 3 / N / PE; 230 / 400 V, 3 / N / PE; 240 / 415 V
Nominal AC voltage range	160 V – 280 V	160 V – 280 V
AC power frequency / range	50 Hz, 60 Hz / -6 Hz ... +5 Hz	50 Hz, 60 Hz / -6 Hz ... +5 Hz
Rated power frequency / rated grid voltage	50 Hz / 230 V	50 Hz / 230 V
Max. output current	16 A	19.2 A
Power factor at rated power	1	1
Adjustable displacement power factor	0.8 overexcited ... 0.8 underexcited	0.8 overexcited ... 0.8 underexcited
Feed-in phases / connection phases	3 / 3	3 / 3
Efficiency		
Max. efficiency / European weighted efficiency	98.1 % / 97.7 %	98.1 % / 97.7 %
Protective devices		
DC disconnect device	●	●
Ground fault monitoring / grid monitoring	● / ●	● / ●
DC surge arrester type II	○	○
DC reverse polarity protection / AC short-circuit current capability / galvanically isolated	● / ● / –	● / ● / –
All-pole-sensitive residual-current monitoring unit	●	●
Protection class (according to IEC 62103) / overvoltage category	I / III	I / III
General data		
Dimensions (W / H / D)	665 / 690 / 265 mm (26.2 / 27.2 / 10.4 inch)	665 / 690 / 265 mm (26.2 / 27.2 / 10.4 inch)
Weight	59 kg / 130.07 lb	59 kg / 130.07 lb
Operating temperature range	-25 °C ... +60 °C / -13 °F ... +140 °F	-25 °C ... +60 °C / -13 °F ... +140 °F
Noise emission (typical)	51 dB(A)	51 dB(A)
Self-consumption (night)	1 W	1 W
Topology / cooling concept	Transformerless / OptiCool	Transformerless / OptiCool
Degree of protection / connection area degree of protection	IP65 / IP54	IP65 / IP54
Climatic category (according to IEC 60721-3-4)	4K4H	4K4H
Max. permissible value for relative humidity (non-condensing)	100 %	100 %

Figure 3.22: *Inverter Name plate and its Datasheet*

III. Array Mounting Structure

Support structures and module mounting arrangements should comply with –

- Applicable building codes regulations
- Standards and module manufacturer's mounting requirements

The following aspects are to be considered in designing array mounting structure:

- Thermal aspects – to allow expansion/contraction of modules/structure
- Mechanical loads on PV structures – to comply with related standards
- Wind – shall be rated for the maximum expected wind speeds
- Material accumulation on PV array – Snow, ice, or other material
- Corrosion – mounting shall be made from corrosion-resistant materials suitable for the lifetime and duty of the system

RCC Roof	Roof Integrated	Superstructure

Figure 3.23: *Different rooftop array mounting structures*
(http://www.ezysolare.com/blog/knowledge-center/cost-of-setting-up-a-rooftop-solar-plant/)

IV. Balance of System Components

The balance of system (BoS) equipment must be selected and installed correctly. If appropriate design procedures are not followed through, the system could have performance and reliability problems, premature faults, and even failure.

The key balance of system components include:

- DC cables and AC cables
- Array junction box/DC combiner box

- Disconnection devices (Over current protection (OCP) device/circuit breakers)
- Plugs, sockets, and connectors
- Energy meters
- Lightning protection system
- Earthing and bonding arrangement
- System monitoring
- Marking and signage

DC cables are used to connect the modules to the array junction box, the array junction box to the DC isolator, and the DC isolator to the inverter. It is important to select a cable that meets the requirements of Draft IEC 62548: Design requirement for PV arrays (in terms of voltage and current specifications) and minimises voltage drop.

AC cables connect the inverter to distribution board and AC loads. The voltage from the inverter is typically 230V AC single phase and the cables required are the same cables used in electrical installations for general household/building cabling. In larger systems, the inverters maybe 400 V AC three phase.

Array Junction Box/DC combiner box is used when the array is comprised of a number of parallel strings. Cables from the array strings are interconnected in an array junction box.

If there are multiple parallel strings, the array junction box will house the connection of the positive and negative cables and fuses from the different PV strings to identified links (or similar). There will be only one DC positive and negative array cable from the junction box interconnecting with the inverter (via the PV array DC isolator)

The PV arrays and string combiner boxes shall at least be IP 54 compliant in accordance with IEC 60529 and shall be UV resistant.

Figure 3.24: *Array Junction Box (http://henselelectric.blogspot.in/)*

Overcurrent protection device or circuit breakers PV modules are current-limited sources but can be subjected to over currents. The over current can originate from multiple parallel strings or inflows from external sources.

Overcurrent within a PV array can also result from earth faults in array wiring or from fault currents due to short circuits in modules, in junction boxes, combiner boxes, or in module wiring.

Disconnection Devices such as fuses and isolators are used to protect the system during the time of maintenance.

Plugs, sockets, and connectors for use at the array level should be rated to DC use. They should be also rated for outdoor use with the appropriate UV resistant and IP rating. The plugs and sockets used for household equipment should not be used in PV arrays.

Lightning protection systems (LPS) are installed in areas where there is high occurrence of thunderstorms. Most buildings are already equipped with LPS and in such cases, proper integration with the RTPV system is required as per IEC 62305 – 3.

Earthing and bonding arrangement for earthing exposed conductive parts of a PV array arises for the following reasons:

- Bonding to avoid uneven potentials across an installation,
- Protective earthing so that there is no faulty current path
- To provide lightning protection

In order to perform these operations, an earthing conductor must be used as per the guidelines prescribed in IEC 62548.

Energy meters record the electrical energy consumed by the loads in kWh within the building where the meter is connected. The electricity consumer is then billed for this consumption on the tariff set for that consumer. Electricity distributors will often have different rates for residential houses compared with industrial and/or commercial consumer.

There are two categories of energy meters in the context of solar PV installation that can be seen in Figures 3.25 and 3.26.

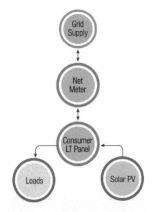

Single Bi-directional meter for net metering

Two Uni-directional meters for net metering

Figure 3.25: *Solar meter*

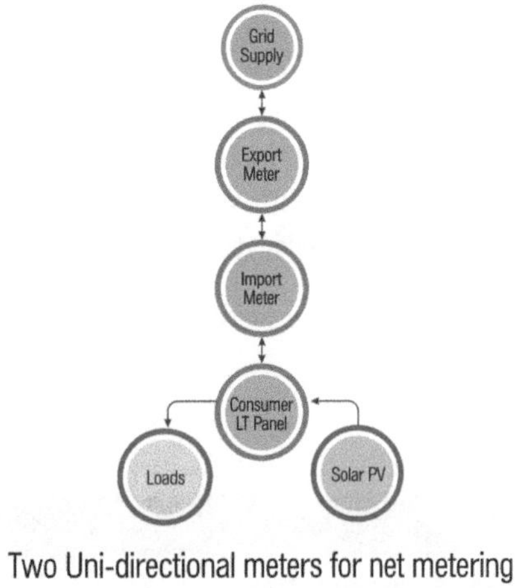

Figure 3.26: *Bidirectional net meter*

BOS Components–Quality and Standards (Adopted by MNRE)

Solar PV Mounting Structure

- IS 2062/ IS 4759: material for mounting structures

Fuses

- IEC 60947: General safety requirements for connectors, switches, circuit breakers
- IEC 60269: Low voltage fuses

Cables

- IEC 60227 & IEC 60502: General testing and measuring method for cables
- BS EN 50618: For DC cables

Surge Arrestors

- IEC 61643-11:2011 Low voltage surge protective devices

Earthing/ Lightning

- IEC 62561 (1,2 & 7): Requirements for connection components, conductors, and earth electrodes and earthing enhancing compounds

Junctions Boxes

- IEC 60529: IP 65 and IP 54 type junction boxes

Energy Meter

- IS 16444: Smart meter Class 1 and 2

System monitoring plays an important role as it gives an overview of the entire system based on how many units of kWh are generated per annum.

Most inverters are equipped with a data module and communication system that constantly updates the amount of power generated on a timely basis (hourly, daily, monthly) and sends it to its respective online portal on which the customer can monitor his/her system.

For commercial plants, system monitoring is more relevant as the investment is large scale.

Figure 3.27: System Monitoring application (https://www.sunnyportal.com)

3.4 KEY TECHNICAL CONSIDERATIONS, STANDARDS, AND SPECIFICATIONS

This section discusses key technical considerations from an administrative stakeholder's perspective, especially for the Discom, in terms of safety, quality, and performance. Discoms should ensure compliance of these factors for PV systems connecting to the distribution grid through appropriate standards and specifications indicated here.

CEA's (Technical Standards for Connectivity of the Distributed Generation Resources) Regulations, 2013 primarily govern the standards and guidelines for rooftop PV systems in India. These regulations refer to relevant IS issued by the Bureau of Indian Standards (BIS). Further, in case of absence of relevant IS, equivalent international standard should be followed in the following order: (a) IEC, (b) British Standard

(BS), (c) American National Standard Institute (ANSI), or (d) any other equivalent international standard. The regulations also state that industry best practices for installation, operation, and maintenance should also be followed along with the relevant standards.

- **IEC 60364, 1ˢᵗ Ed. (2002-05), 'Electrical installations of buildings – Part**

  7-712: Requirements for special installations or locations –PV power supply systems,' is the primary standard for PV installations, safety and fault protection, common rules regarding wiring, isolation, earthing, etc. This standard is applicable and commonly followed in India. This standard is also equivalent to and/ or in conjunction with other standards around the world such as:

  o DIN VDE 0100-712:2006-06, Part 7-712: Requirements for Special Installations or Locations Solar photovoltaic (PV) Power Supply Systems.

  o UL 1741: Standard for Inverters, Converters, Controllers, and Interconnection System Equipment for Use with Distributed Energy Resources.

  o IEEE 1547: Standard for Interconnecting Distributed Resources with Electric Power Systems.

  o IEEE 929-2000: Recommended Practice for Utility Interface of Photovoltaic (PV) Systems.

  o NEC 690: Solar Photovoltaic (PV) Systems.

What Is 'Anti-Islanding?'

One of the foremost concerns among DISCOMs (and even transmission companies) engineers, when connecting a PV system to the grid is *'What if the distribution grid shuts down but the PV system remains "on" and keeps on injecting power into the grid? Could this be a hazard to the technician who is unaware of this live PV system and comes in direct physical contact with the grid?'*

Another common question is *'If two PV systems are feeding solar power into the grid, and if the grid shuts down, can the two inverters create a reference for each other and remain on?'*

The answer to both these questions is 'NO.' The good news is that this problem has been sorted out a long time ago and is successfully being practiced around the world.

All grid-connected PV inverters are designed to shut down when grid parameter changes beyond the predefined range programmed in the inverter (including grid shut-down); thus, avoiding the PV system to act as an energised 'island.' This feature is called anti-islanding.

Anti-islanding is ensured through various IEEE, IEC, UL, DIN VDE, standards for such grid-connected inverters.

a. **Electrical Safety**

i. <u>General</u>: All PV systems should comply with the CEA's (Measures Relating to Safety and Electricity Supply) Regulations, 2010.

ii. Anti-Islanding: All grid-connected and hybrid PV inverters are designed to shut-down when the grid parameters like voltage, frequency, rate of change of frequency, etc. change beyond the predefined range of the inverter.

- **IEC 61727, 2ndEd. (2004), 'Photovoltaic (PV) systems – Characteristics of the utility interface,' is a standard for PV systems rated for 10 kVA or less. Section 5.2.1 indicates maximum trip time in response to grid voltage variation as given in Table 3.2. Section 5.2.2 specifies that the system should cease to energise the grid within 0.2 seconds if the grid frequency deviates beyond ±1 Hz of nominal frequency.**

- **CEA's (Technical Standards for Connectivity of the Distributed Generation Resources) Regulations, 2013, in its Section 11 (6) stipulates similar response times for disconnection of the distributed generation system. However, IEC 61727 being more stringent as well as widespread is acceptable and more convenient to follow in India.**

- **IEC 62116, 2nd Ed. (2014-02), 'Utility-interconnected photovoltaic inverters – Test procedure for islanding prevention measures,' provides a test procedure to evaluate the performance of islanding prevention measures for inverters that are connected to the utility grid. Inverters complying with this standard, for capacities both less than and greater than 10 kVA, are considered non-islanding as defined in IEC 61727.**

iii. <u>Earthing (or Grounding)</u>:

While earthing practices in India are common and guided by IS: 3043-1987 (Reaffirmed 2006), although a PV system contains both AC and DC equipment, earthing practices are often not obvious for such systems. Hence, clarification regarding earthing practices become critical from the System Designer's as well as the Electrical Inspector's perspective.

- **IS 3043-1987 (Reaffirmed 2006), 'Code of practice for earthing,' governs the earthing practices of a PV system.**

Table 3.2: Trip Time in Response to Abnormal Voltages as per IEC 61727

Grid Voltage (at Interconnection)		Maximum Trip Time
V < 50% of $V_{Nominal}$	:	0.1 seconds
50% < V < 85%	:	2.0 seconds
85% < V < 110%	:	Continuous
110% < V < 135%	:	2.0 seconds
135< V	:	0.05 seconds

Note: $V_{Nominal}$ for India is 240 V (1φ) or 415 (3φ) V at LT.

Earthing is required for PV module frames, array structures (power, communication, and protective), equipment and enclosures, AC conductors, and lightning conductors. Although DC and AC systems are considered separate, they should be connected together during earthing.

Earthing of DC cable is not required in most cases. However, some inverters (usually with transformers) allow DC conductor earthing. In such cases, if allowed by the inverter, the negative DC cable should be connected to earth in order to reduce Potential-Induced Degradation (PID) of the PV modules. PID and methods to mitigate it are discussed in latter sections of this chapter.

Only earthing of the lightning conductor should be isolated from the earthing of the remaining PV system.

All inverters should have provision for earth fault monitoring and shall disconnect from the grid and shut down in case of earth faults. The IEC 62109-2nd standard includes earth fault protection requirement for PV circuits.

- **IEC 62109-1, 1st Ed. (2010-04), 'Safety of power converters for use in photovoltaic power systems – Part 1: General requirements,' defines the minimum requirements for the design and manufacture of Power Conversion Equipment (PCE) for protection against electric shock, energy, fire, mechanical, and other hazards.**

- **IEC 62109-2, 1st Ed. (2011-06), 'Safety of power converters for use in photovoltaic power systems – Part 2: Particular requirements for inverters,' defines the particular safety requirements relevant to DC to AC inverter products as well as products that have or perform inverter functions in addition to other functions, where the inverter is intended for use in photovoltaic power systems. Inverters covered by this standard may be grid-interactive, stand alone, or multiple mode inverters may be supplied by single or multiple photovoltaic modules grouped in various array configurations and may be intended for use in conjunction with batteries or other forms of energy storage. This standard must be used jointly with IEC 62109-1.**

When earthing PV modules, all frames should be connected to one continuous earthing cable. Many installers use small pieces jumper cables to connect frames of consecutive modules, which is a wrong practice. Further, star-type washers should be used when bolting the lugs of earthing cable with the module frame that can scratch the anodisation of the module frame to make contact with its aluminium.

The earthing conductor should be rated for 1.56 times the maximum short circuit current of the PV array. The factor 1.56 considers 25% as a safety factor and 25% as albedo factor to protect from any unaccounted external reflection onto the PV modules increasing its current.

In any case, the cross-section area or the earthing conductor for PV equipment should not be less than 6 mm2 if copper, 10 mm2 if aluminium, or 70 mm2 if hot-dipped galvanised iron. For the earthing of lightning arrestor, cross-section of the earthing conductor should not be less than 16 mm2 of copper or 70 mm2 if hot-dipped galvanised iron.

Resistance between any point of the PV system and earth should not be greater than 5 Ω at any time. All earthing paths should be created using two parallel earth pits to protect the PV system against failure of one earth pit.

(iv) **DC Overcurrent Protection**: As the output current of the PV module is limited by the amount of sunlight received, the maximum current on the DC side of the PV system is calculated based on the rated short circuit current of the PV module.

The PV system is protected from overcurrent from the PV modules with the help of fuses at the string junction box. As PV module are connected in series in a string, the short circuit current of the string is equal to the short circuit current of the PV module. Each string should have two fuses, one connected to the positive and the other to the negative terminal of the string. The fuse should be rated at 156% of short-circuit current and 1,000 VDC; if the exact current rating is not available, the nearest available higher rating should be used. However, the rating of the fuse should not exceed 200% of the short circuit current of the string. The fuse should be housed with dedicated fuse disconnectors.

DC MCB is an alternative option to fuses. They also provide an added advantage of allowing isolation of individual strings. However, this is a more expensive option compared to fuses, and there are also chances of accidental tripping of the MCB.

(v) **DC Surge Protection**: Several makes for DC surge arresters or SPD are available specifically for PV applications. The surge arrestors should be of Type 2 (with reference to Standard IEC 61643-1, 'Low Voltage Surge Protective Devices'), rated at a continuous operating voltage of at least 125% of the open-circuit voltage of the PV string and a flash current of more than 5 A. As the string inverters used for rooftop PV systems do not allow more than 800 VDC, surge arrestors rated for 1,000 VDC are commonly used. The surge arrestors should be connected to both positive and negative outgoing terminal of the string junction box (if the inverter already does not have an equivalent in-build DC surge arrestor).

(vi) **Lightning Protection**: Lightning can cause damage to a PV system either by a direct strike or through surge in the grid resulting from a nearby lightning strike.

Lightning protection installations should follow IS 2309-1989 (Reaffirmed 2010).

- **IS 2309-1989 (Reaffirmed 2010), 'Code of practice for the protection of buildings and allied structures against lightning' governs all lightning protection-related practices of a PV system.**

Small rooftop PV systems pose minimal risk of lightning strike, and the cost impact of lightning protection system can be substantial. Hence, it may not be required to have a lightning protection system for rooftop PV systems of capacity less than 10 kW.

It is recommended for larger PV systems to have a dedicated lightning protection system including lightning rods, conductors, and dedicated earth pits. Already existing lightning protection of a building may be used, provided it adequately protects the installation area and is assured of functioning throughout the life of the PV system.

(vii) **Ingress Protection**: All PV equipment, if installed outdoors, should have an ingress protection rating of at least IP65. This strictly applies to all junction boxes, inverters, and connectors. Although many inverters are rated for operation up to a maximum ambient temperature of 60°C, it is highly recommended to make an additional shading arrangement to avoid exposure to direct sunlight and rain.

(viii) **Labelling of PV System Equipment**: Labelling of PV equipment is a crucial aspect of safety owing to the high DC voltages as well as non-familiarity of technicians and laymen with such a system. The labelling of a PV system should conform to IEC 62446 standard.

- **IEC 62446, 1st Ed. (2009-05), 'Grid-connected photovoltaic systems – Minimum requirements for system documentation, commissioning tests and inspection,' defines the minimal information and documentation required to be handed over to a customer following the installation of a grid-connected PV system. This standard also describes the minimum commissioning tests, inspection criteria, and documentation expected to verify the safe installation and correct operation of the system.**

IEC 62446 stipulates that:

- All circuits, protective devices, switches, and terminals are suitably labelled.
- All DC junction boxes (PV generator and PV array boxes) carry a warning label indicating that active parts inside the boxes are fed from a PV array and may still be alive after isolation from the PV inverter and public supply.
- The main AC isolating switch is clearly labelled.
- Dual supply warning labels are fitted at points of interconnection.

- A single line wiring diagram is displayed on site.
- Inverter protection settings and installer details are displayed on site.
- Emergency shutdown procedures are displayed on site.
- All signs and labels are suitably affixed and made durable.

b. **Electrical Quality:**

(i) **DC Power Injection**: Most grid-connected inverters are transformer-less and hence utilities are concerned about DC power injection into the grid. DC power injection is restricted to either an absolute value or a minor fraction of the rated inverter output current.

- **IEC 61727, 2nd Ed. (2004), 'Photovoltaic (PV) systems – characteristics of the utility interface,' in Section 4.4 stipulates that the PV system shall not inject DC current greater than 1% of the inverter-rated output current into the grid.**

- **CEA's (Technical Standards for Connectivity of the Distributed Generation Resources) Regulations, 2010 in its Section 11 (2) stipulates that the distributed generating resource shall not inject DC greater than 0.5%` of the full-rated output at the interconnection point.**

(ii) **Harmonic Injection**: Most inverters in India are rated for THD of less than 3% of power injected into the grid and hence are suitable for interconnection from a harmonic injection standpoint.

- **CEA's (Technical Standards for Connectivity of the Distributed Generation Resources) Regulations, 2010 in its Section 11 (1) stipulate that harmonic current injections from a generating station shall not exceed the limits specified in (Standard) IEEE 519.**

- **IEEE 519 (2014), 'Recommended practice and requirements for harmonic control in electric power systems,' stipulates the voltage and current harmonic injection limits as indicated in Table 3.3 and Table 3.4, respectively.**

(iii) Phase Imbalance (or Unbalance): Phase imbalance can occur due to varied loads and power injected into different phases of the distribution grid. The Discom should always limit its voltage imbalance to less than 3%.

Phase imbalance can potentially arise from single-phase inverters feeding into the distribution grid. It is typically observed that multiple rooftop PV interconnections tend to have an averaging effect on the grid and do not pose substantial unbalancing threats. However, it is recommended that Discoms should keep track of the PV capacity connected to each phase for troubleshooting any extreme cases.

Three-phase inverters (and PV systems) rather aid in minimising the phase imbalance as they tend to uniformly feed power into all three phases.

Table 3.3: Voltage Distortion Limits as per IEEE 519 (2014)

Bus Voltage (V) at PCC	Individual Harmonic	Total Harmonic Distortion (THD)
V< 1.0 kV	5.0 %	8.0 %
1 kV <V< 69 kV	3.0 %	5.0 %
69 kV <V< 161 kV	1.5 %	2.5 %
161 kV <V	1.0 %	1.5 %*

Notes: PCC: Point of Common Coupling.
*High-voltage systems can have up to 2.0 % THD where the cause is an HVDC terminal whose effects will have attenuated at points in the network where future users may be connected.)

Table 3.4: Current Distortion Limits as per IEEE 519 (2014)

Maximum Harmonic Current Distortion in Percentage of IL						
Individual Harmonic Order (Odd Harmonic)						
ISC/IL	<11	11<h<17	17<h<23	12<h<35	35<h	TDD
<20*	4.0	2.0	1.5	0.6	0.3	5.0
20<50	7.0	3.5	2.5	1.0	0.5	8.0
50<100	10.0	4.5	4.0	1.5	0.7	12.0
100<1,000	12.0	5.5	5.0	2.0	1.0	15.5
>1,000	15.0	7.0	6.0	2.5	1.4	20.0

Notes: Even harmonics are limited to 25 % of the odd harmonic limits.
Total demand distortion (TDD) is based on the average maximum demand current at the fundamental frequency, taken at the point of common coupling (PCC).
*All power generation equipment is limited to these values of current distortion regardless of ISC/IL.
ISC: Maximum short circuit current at the PCC.
IL: Maximum demand load current (Fundamental) at the PCC.
h: Harmonic number.

(iv) Flicker: IEC 61000 is a set of standards on electromagnetic compatibility which are subdivided into sections that define:

- The environment from the EMC viewpoint and establish the compatibility levels that the distributors must guarantee.
- The emission levels into the networks.
- The immunity levels of the appliances.

Table 3.5: IEC Standards and Scope for Electromagnetic Compatibility, including Flicker

Standard	Subject	Scope
IEC 61000-6-1	Immunity	Residential and Commercial
IEC 61000-6-3	Emission	
IEC 61000-6-2	Immunity	Industrial
IEC 61000-6-4	Emission	
IEC 61000-3-2	Harmonics	**Inverter < 16 A AC**
IEC 61000-3-3	**Voltage Fluctuation and Flicker**	**Current per Phase**
IEC 61000-3-12	Harmonics	**Inverter > 16 A and < 75 A**
IEC 61000-3-11	**Voltage Fluctuation and Flicker**	**AC Current per Phase**
IEC 61000-3-4	Harmonics	**Inverter > 75 A AC Current**
IEC 61000-3-5	**Voltage Fluctuation and Flicker**	**per Phase**

The relevant IEC 61000 sections for electromagnetic compatibility including voltage fluctuation and flicker are shown as follows:

- **CEA's (Technical Standards for Connectivity of the Distributed Generation Resources) Regulations, 2010 in its Section 11 (3) stipulate that distributed generating resource shall not introduce flicker beyond the limits specified in IEC 61000.**
- **IEC 61727, 2nd Ed. (2004), 'Photovoltaic (PV) systems – characteristics of the utility interface,' in Section 4.3 stipulates that the operation of the PV system should not cause voltage flicker in excess limits stated in the relevant sections of IEC 61000-3-3 for systems less than 16 A or IEC 61000-3-5 for systems with the current of 16 A and above.**

(v) **Power Factor**: Grid-connected PV inverters are typically capable of injecting energy into the grid at unity power factor and hence tend to have a positive impact on the grid.

Many inverters also provide a power factor range such as −0.8 (inductive) to +0.8 (capacitive), which can be pre-programmed or can be dynamically adjustable.

c. **PV Module Considerations:**

As the PV module is the most expensive component of the PV system, it is extremely critical to outline its specification. However, on the positive side, the PV module is a very robust component and hence satisfactory quality and performance can be ensured by checking key standards and specifications. One must also be sensitised that over-specification of the PV module can result into a substantial cost increase without any major gain in quality or performance of the module.

(i) **Components of a PV Module:** The main components of a polycrystalline silicon PV module are:

- *Solar Cell or PV Cell is a device that directly converts sunlight into electricity. A standard polycrystalline silicon PV module has 60 or 72 solar cells connected in series. The area of each solar cell is 156 mm x 156 mm (6 inch x 6 inch) and gives an output current of around 8 A$_{DC}$ and voltage of around 0.5 V$_{DC}$, resulting into an output power of approximately 4 Watts at STC. It should be noted that the mentioned values of current, voltage, and power vary from cell to cell and hence PV modules of varying ratings exist in the market. All cells in a PV module are connected in series.*

- *Bus Ribbons are soldered to the positive terminal of one solar cell and the negative terminal of the other solar cell, thus, electrically connecting them in series.*

STC and PV Module Efficiency

STC implies a solar spectrum of Air Mass (AM) 1.5 at 1,000 watts per square meter perpendicularly incident on a PV module, wherein the PV module temperature is fixed at 25°C. (Simply stated, AM 1.5 implies a representative solar radiation and spectrum experienced on a typical sunny day on the earth's surface.)

All PV modules are tested for their electrical outputs at STC using a solar simulator at the time of manufacturing. The resultant output power is called the 'rating' of that PV module and denoted in watt (W) or watt-peak (Wp).

Hence, if a PV module is rated for 250 W or 250 Wp, then it will give an output of 250 W on the noon of a sunny day, if the PV module is facing the sun and the temperature of the module is 25°C. However, as the sun moves relatively to the PV module from this position and/or the temperature of the PV module increases, then the power output of the PV module will also decrease.

If the area of the 250 Wp PV module is 1 m x 1.6 m, then the efficiency (η) of that module is calculated as follows:

$$\text{Efficiency} = \frac{250W}{1000 \frac{W}{m^2} \times 1.6m^2)}$$

$$= 15.63\%$$

- *Glass provides mechanical strength to the PV module and also protects the internal components from the external harsh environment. This glass is tempered, low-iron, high in transmissivity and is 3-4 mm thick.*

- *Ethylene Vinyl Acetate (EVA), a transparent thermoplastic, is used above and below the solar cells to encapsulate them. EVA protects the solar cells from mechanical shocks while allowing the light to transmit through it.*

- *Tedlar® or a similar back-sheet is used to protect the PV module from radiation, moisture, and weather while also providing electrical insulation to the module. Some PV modules, especially for building integration or aesthetic purposes, may use a second glass layer instead of a back-sheet.*

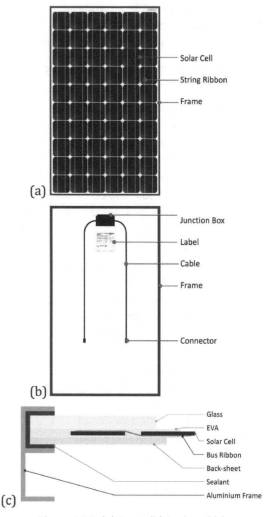

Figure 3.28: *(a) Front, (b) Back and (c) Cross-Sectional View of a PV Module*

- *Junction box affixed at the back-side of the PV module connects the internal conductors of the module to external cables for connection. Junction boxes also contain bypass diodes to provide an alternate path for current in case certain sections of the PV module are not able to conduct or generate power due to shading or damage.*

- *Cables and connectors are an integral part of the PV module and have to comply with the general standards of DC cables and connectors.*

- *Edge sealant may be a silicone compound or a tape, and is used to protect the PV module from moisture and dust ingress from the sides and also to hold the frame.*

- *Frames are typically made of anodised aluminium and are used to protect the PV module, mount the module using clamps or bolts, and connect to the body earthing of the overall module.*

(ii) **Rating of a PV Module:** A PV module is rated for its power output at STC. The PV module is also rated for its open-circuit voltage (VOC); short-circuit current (ISC), voltage at maximum power point (VMP), and current at maximum power point (IMP). In addition to these, the temperature coefficient of power, voltage, and current are also indicated in the datasheet of the PV module, which is important for designing as well as estimating the output of the PV system.

(iii) **Basic Design and Safety Qualification:** The design of the PV module is guided by one of the following three IEC standards depending on the type of the module, i.e., IEC 61215 for crystalline silicon, IEC 61646 for thin-film, or IEC 62108 for concentrator PV modules.

- **IEC 61215, 2nd Ed. (2005-04), 'Crystalline silicon terrestrial photovoltaic (PV) modules – Design qualification and type approval,'** outlines all the procedures for sampling, marking, and testing of mono- and multi-crystalline silicon PV modules. The testing includes visual inspection, maximum power determination, insulation test, measurement of temperature coefficients, measurement of nominal operating cell temperature (NOCT), performance at STC and NOCT, performance at low irradiance, outdoor exposure test, hot-spot endurance test, UV pre-conditioning test, thermal cycling test, humidity-freeze test, damp-heat test, robustness of terminations test, wet leakage current test, mechanical load test, hail test, and bypass diode thermal test.

- **IEC 61646, 2nd Ed. (2008-05), 'Thin-film terrestrial photovoltaic (PV) modules – Design qualification and type approval,'** outlines all the procedures for sampling, marking, and testing of thin-film PV modules such as amorphous silicon, cadmium telluride (CdTe), copper indium gallium selenide (CIGS), micromorph, and similar technologies. The testing includes visual inspection, maximum power determination, insulation test, measurement of temperature coefficients, measurement of nominal operating cell temperature (NOCT), performance at STC and NOCT, performance at low irradiance, outdoor exposure test, hot-spot endurance test, UV pre-conditioning test, thermal cycling test, humidity-freeze test, damp-heat test, robustness of terminations test, wet leakage current test, mechanical load test, hail test, bypass diode thermal test, and light soaking.

- **IEC 62108, 1st Ed. (2007-12), "Concentrator photovoltaic (CPV) modules and assemblies – Design qualification and type approval,"** outlines all the procedures for sampling, marking and testing of concentrator cell technologies and assemblies. The testing includes visual inspection, electrical performance measurement, ground path continuity test, electrical insulation test, wet insulation test, thermal cycling test, damp heat test, humidity freeze test, hail impact test, water spray test, bypass/blocking diode thermal test, robustness of termination test, mechanical load test, off-axis beam damage test, ultraviolet conditioning test, outdoor exposure test, and hot-spot endurance test.

In addition to one of the above-mentioned three IEC certifications, all PV modules should also be certified for IEC 61730 as a part of their safety qualification.

- IEC 61730-1, Ed. 1.2 (2013-03), 'Photovoltaic (PV) module safety qualification – Part 1: Requirements for construction,' describes the fundamental construction requirements for PV modules in order to provide safe electrical and mechanical operation during their expected lifetime. Specific topics are provided to assess the prevention of electrical shock, fire hazards, and personal injury due to mechanical and environmental stresses. This part pertains to the particular requirements of construction.

- IEC 61730-2, Ed. 1.1 (2012-11), 'Photovoltaic (PV) module safety qualification – Part 2: Requirements for testing,' describes the fundamental construction requirements for PV modules in order to provide safe electrical and mechanical operation during their expected lifetime. Specific topics are provided to assess the prevention of electrical shock, fire hazards, and personal injury due to mechanical and environmental stresses. This part pertains to the particular requirements of testing.

One or both IEC certifications may be applicable if PV modules are intended for continuous outdoor exposure to highly corrosive wet environments.

- IEC 61701, 2nd Ed. (2011-12), 'Salt mist corrosion testing of photovoltaic (PV) modules,' describes the test sequence useful to determine the resistance of different PV modules corrosion from salt mist containing Cl- (NaCl, MgCl2, etc.). This standard is applicable to PV modules intended for continuous outdoor exposure to highly corrosive wet environments such as marine environments or temporary corrosive environments where salt is used in winter periods to melt ice formations on roads.

- IEC 62716, 1stEd. (2013-06), 'Photovoltaic (PV) modules – ammonia corrosion testing,' describes the test sequence useful to determine the resistance of different PV modules to ammonia (NH3). This standard is applicable to PV modules intended for continuous outdoor exposure to wet atmospheres having high concentration of dissolved ammonia such as stables of agricultural companies.

(iv) **Performance Warranty:** The performance warranty of a PV module is one of the most critical considerations while procuring the module. The globally accepted performance warranty commits less than 10 % performance degradation in power output during the first 10 years and less than 20 % performance degradation during the subsequent 15 years. Tier-I module manufacturers also back their performance warranty with bank guarantees as an added assurance.

(v) **Workmanship Warranty:** The typical workmanship warranty on a PV module is five years.

(vi) **Potential-Induced Degradation (PID):** The high potential difference between the solar cell and module frame (which is grounded) drives ion mobility between them, which is further accelerated by humidity and temperature; all these phenomena causes degradation in the output power of the PV module.

Hence, it is recommended to use PID-resistant PV modules which resist the transportation of causative ions such as Na+ leaking from the glass, EVA, or even the anti-reflective coating of the solar cell. PID-resistant modules use highly insulating EVA with Na+ blocking capabilities, low conducting glass cover, higher distances of the cell strings to the frame, insulating frame, and so on.

PID can also be eliminated by grounding the negative terminal of the PV string if inverters with transformers are used. Alternatively, PID can also be eliminated or reversed via the application of a reverse voltage (using an external power supply, often called 'PID-box') during night time to the module strings or to specific modules.

d. **Mechanical and Workmanship Considerations:**

(i) **Inclination of PV Modules**: The optimal angle of inclination of a flat plate solar collector (which also includes a fixed PV module) is very close to the latitude of the location of installation. Further, the PV modules installed in the northern hemisphere (as is the case for India) should be inclined such that they face south.

However, it is also a common practice to reduce the inclination angle in the range of 10–15° (irrespective of the latitude of location) on flat roofs or terraces. Such reduction in inclination results into simpler, quick, and cost-effective installation owing to lesser wind resistance of the low profile, lighter mounting structure, and avoided penetration or anchoring into the terrace. The reduction in solar energy generation can be relatively up to 5%, but the commercial and other benefits tend to outweigh this loss.

(ii) **Area of a Rooftop PV System**: A rooftop PV system can take anywhere from 10 to 15 m2 of area per kilowatt of installation depending on the angle of inclination of the PV modules. This area also includes the spacing between two rows of PV modules.

(iii) **Weight of the Rooftop PV System**: The weight of a PV system (including the PV module and structure) does not exceed 30 kg per m2. However, for mounting structures that are not anchored into the roof the weight of the PV system is deliberately increased using bricks to counter the uplift or drag forces created by wind pressures. In any case, all terraces are designed to withstand the weight of PV systems.

(iv) **Wind Loads:** All MMS should be designed taking into consideration the wind loads at the location of installation. The design should consider the 'wind speed zone' of the location as per Indian Standard IS 875 (Part 3)-1987.

- **IS 875 (Part 3)-1987, 'Code for practice of design loads (other than earthquake) for buildings and structures,' guides the design principles of wind loads to be considered when designing buildings, structures, and its components. This standard is directly applicable to the design of PV module mounting structures.**

The design document of a module mounting structure is a mandatory component of the overall design of the rooftop PV system. This design should be developed or approved by a chartered structural engineer. This design should also be a part of the submission for drawing and design approval to the concerned electrical inspector or inspection agency.

Readymade and modular mounting structures pre-certified for certain wind speeds and are readily available in the market, and the same can be directly used.

For PV installations on tall buildings, the design should consider the 'height factor' as per IS 875 (Part 3)-1987, which quantifies higher wind loads on tall structures within the same wind zone.

(v) **Material of Mounting Structure**: Galvanised iron (GI) or aluminium is the most common material used for module mounting structures. In case of GI structures, the quality of galvanisation becomes very critical to ensure a rust-free life of at least 25 years.

It is highly recommended to use stainless steel fasteners due to their weather-resistant properties. If stainless steel is not possible due to any reason, then GI fasteners can be used.

(vi) **Penetration and Puncturing of Roof or Terrace:** Penetration into or puncturing the roof or terrace for anchoring of MMS should be avoided as far as possible to avoid any water leakage-related issues.

However, if puncturing the roof is unavoidable, sufficient care should be taken for waterproofing the roof or terrace as a part of the installation itself.

e. **Other Considerations:**

(i) Performance of a PV System: The quantum of energy output of a PV system depends on:

- System properties such as its capacity, internal losses, and tracking (if used), maintenance practices and frequency of cleaning.

- Weather parameters such as incident radiation and temperature as well as ambient factors like fog and pollution.

- Gird parameters such as fluctuations in voltage and frequency and availability.

3.5 POLICY AND REGULATION

3.5.1 Key Stakeholders

Rooftop solar PV programs in all states are guided by the particular state's policies, regulations, and guidelines by Discoms.

Most states strive to adopt a single window clearance that helps consumers and investors obtain all clearances at a single office. This must be implemented in true spirit, and a dedicated team may be constituted for the rapid approvals and addressing of investor's grievances. Many states are at various stages on this and utility engineers must check their own state's policy and regulation for rooftop PV deployment so as to be conversant with the framework of the same.

There can be different stakeholders in such programme and following are the key stakeholders as far as permitting framework for rooftop PV is concerned.

a. **State Nodal Agency**

A nodal agency is a Government department that is responsible for promotion of the policy. Clear demarcation of responsibility and a single point of contact for potential consumers/investors go a long way in improving the overall investment climate of the state.

The role of nodal agency may change from state to state but generally it gets involved in implementing policies and in permitting subsidies, wherever available by Central or state government. It can also play a role in enlisting state-level installers or system integrators for grid-connected rooftop systems, which can be referred to by Discoms to allow the consumers to choose from. The nodal agency should also be taking active role in confirming necessary quality in rooftop installations and components. It can take reference from MNRE-approved standards and technical specifications.

b. **Implementing Agency – The Discom**

While the nodal agency will be responsible for promotion of the policy and passing on benefits (e.g., subsidies) to stakeholders, it is the implementing agency that is responsible for implementing the solar rooftop programme.

As grid-connected solar rooftop plants have an implication on utility billing, grid safety, and power quality, the Discom becomes the *de facto* implementing agency.

While, on the other hand, the SNA can become the implementing agency for stand-alone solar projects.

- As of the time of this publication, the nodal agency has some role in implementation of solar rooftop at the state level. However, there can be changes in the implementation framework as per suggestions of MNRE.

c. **Eligible Entities**

Eligible entities are usually among the different categories of electricity consumers mentioned in the SERC orders. State governments may decide to allow all the applicable schemes in the policy to all types of consumers or may choose to limit the schemes to certain consumers due to the financial implication on the state or one of its Discoms. A good example of this is incentives such as banking or net-metering schemes. Such incentives may be restricted or reduced to commercial and industrial consumers while placing no restrictions for residential consumers.

Financial implications on Discoms due to promotion of rooftop solar programme may be considered while setting the targets of the policy. In case the government has financial constraints, the target may correspondingly be reduced to a number that is comfortable to various stakeholders.

The definition of 'eligible entity' becomes more important with evolving solar PV rooftop financing ecosystem with number of new business models. An eligible entity may be an owner of the building, a tenant, or even a third-party investor.

In addition to being a consumer on the Discom's network, the following clause may be present in the policy and/or regulation with regard to eligible entities:

An eligible entity is a:

1. *Person or company that either owns or leases the system from a third-party financer/developer/investor.*
2. *Legal owner of the premise on which the solar rooftop system is to be installed or a tenant of the premises, in case the building is on lease.*
3. *Consumer of the distribution licensee for the area on which the building is located.*

3.5.2 Schemes/Applicable Business Models

A solar rooftop policy is implemented through various schemes. Schemes provide the necessary implementing framework for the policy and may change from time to time within the tenure of the policy. They may also include specific subsidies and incentives that are also time-bound and may be applicable to a certain eligible entities and types of systems (e.g., off-grid or on-grid).

Business models are critical from an investor/company's point of view in order to ensure return on their investment. Revenue for these investors can come in different forms; for example, through a PPA with the rooftop owner/Discom/open access consumer or through direct sale of equipment and engineering services (e.g., an engineering, procurement and construction (EPC) company).

Various incentives and exemptions applicable to a solar PV rooftop programme are summarised in Table 3.6.

Table 3.6: Reference Incentives and Exemptions Applicable to a Solar PV Rooftop Policy

Type of Incentive/ Exemption/ Parameter	Sale to Distribution Licensee		Sale to Third Party
	Net Metering	Gross Metering	Open Access
PV System Capacity	Limited to consumer's contract demand/ sanctioned load	Limited by the available rooftop area (or related to associated distribution transformer capacity) or as per the relevant terms of RFP, if applicable	Based on mutual agreement between developer and off-taker
Ownership	Self-owned	Self-owned or third-party owned	Third-party owned
Demand Cut	50% of the consumer's current billing demand	Not applicable	50% of the consumer's current billing demand
Billing Cycle	As per consumer's current billing cycle	Monthly	Solar energy to be adjusted on a 15-minute basis
Banking	Excess energy allowed to be banked during a financial year, at the end of which excess generation will be paid at an appropriate tariff determined by concerned SERC	Not applicable as complete energy is sold to the distribution licensee at the tariff determined by concerned SERC.	No banking allowed for third party sale of power. Any excess, unadjusted energy shall be purchased by the distribution licensee at the tariff determined by concerned SERC

Type of Incentive/ Exemption/ Parameter	Sale to Distribution Licensee		Sale to Third Party
	Net Metering	Gross Metering	Open Access
Tariff	As determined by SERC from time to time	As determined by SERC from time to time or based on competitive bidding using SERC's tariff as benchmark	Mutually agreed between developer and consumer
Wheeling Charges	Not applicable	Not applicable	As per concerned SERC order
Transmission Charges	Not applicable	Not applicable	As per concerned SERC order
Wheeling Losses	Not applicable	Not applicable	As per concerned SERC order
Transmission Losses	Not applicable	Not applicable	As per concerned SERC order
Cross Subsidy Surcharge (CSS)	Not applicable	Not applicable	As per concerned SERC order
Electricity Duty	Not applicable	Exempted	Exempted
Renewable Energy Certificate (REC)	Consumer can claim REC for solar energy consumed by self and energy sold to distribution licensee at Average Power Procurement Cost (APPC). (In addition, the developer shall abide by all other provision as per the relevant REC regulations.)	Developer can claim REC if selling power to Distribution Licensee at APPC. (In addition, the developer shall abide by all other provision as per the relevant REC regulations.)	Developer can claim REC based on the provisions of relevant REC regulations

Type of Incentive/ Exemption/ Parameter	Sale to Distribution Licensee		Sale to Third Party
	Net Metering	Gross Metering	Open Access
Renewable Purchase Obligation (RPO)	Distribution licensee can claim RPO if (i) consumed solar energy is not credited towards the consumer's RPO and (ii) no REC is claimed for the generated solar energy	Distribution Licensee can claim RPO if no REC is claimed for the generated solar energy	Distribution Licensee can claim RPO if (i) consumed solar energy is not credited towards the consumer's RPO and (ii) no REC is claimed for the generated solar energy
Clean Development Mechanism (CDM)	CDM is retained by the consumer	CDM is retained by the developer	CDM sharing is left to the developer and off taker

3.5.3 Technical Requirements for Interconnection

Technical requirements such as metering and issues concerning grid integration are covered by the CEA standards. There are three specific regulations that are applicable to solar PV rooftop systems:

- The CEA 'Technical Standards for Connectivity of the Distributed Generation Resources' Standards 2013
- The CEA 'Measures Relating to Safety and Electricity Supply' Standards 2010
- The CEA 'Installation and Operation of Meters' 2010

3.5.4 Purpose and Introduction to Regulation

The role of solar PV rooftop regulation is mainly three fold:

1. Determine benchmark capital costs and tariff for solar rooftop grid-connected systems.
2. Specify the grid code, check standards with respect to power quality and other electrical parameters that ensure that the functioning of the grid is not compromised.
3. Ensure the proper interpretation of the Electricity Act, 2003 and resolve any disputes between power producers, Discoms and consumers.

Regulations for solar rooftop systems are only applicable in case of grid-connected systems. The regulations may be either net/gross metering. The Electricity Act, 2003 provides the legal framework for setting up both the Central and the state ERCs in the country.

The Central and state regulators are guided by the National Tariff Plan, National Electricity Policy, and Tariff Policy (Section 79, 86).

In case of any conflicts between the policy and the regulations, the Electricity Act, 2003 (Section 107 and 108) clearly states that the decision of the Central/state government shall be final.

3.5.5 Key Considerations and Components in Framing Regulations

Any net/gross-metering regulation should ideally consider the following clauses:

a. **Title, Scope, and Application**

The regulation should clearly indicate the eligible consumer to whom and under what instances do the regulations apply. This clause is important to enable third-party sale of power via solar rooftop systems. This term is also used in the state/Central solar rooftop policy. The key difference here is that the regulator assesses eligibility on a technical basis such as grid voltages, grid availability. whereas the state/Central government assesses eligibility on other financial and social criteria as well.

b. **Applicable Models**

All models for grid connectivity such as net-metering and gross metering can be addressed here. Although, there is significant overlap with the solar rooftop policy, it is good for the SERC and the state government to be in line with each other on this topic.

There are more states where net metering regulation is announced in comparison with states with gross-metering regulation. The following clause may be present to indicate applicable models:

Only Gross/Only Net/both gross and net metering shall come under the ambit of this regulation.

1. *Gross Metering – Under gross metering, the system owner (consumer of distribution licensee OR third-party financer OR developer) shall export energy into the grid irrespective of the consumption of the building on which the solar PV rooftop system is located. This can be considered as a direct sale to the distribution licensee.*

2. *Net Metering – Under net metering, the consumption of the building on which*

the solar PV rooftop system is installed is set off from the energy exported on to the grid. In this arrangement, the system owner and the consumer of power from the distribution licensee are typically the same. Any financial arrangement between a third-party financer/developer and the rooftop owner are allowed but do not fall under the gambit of this regulation. All agreements between the financer/developer and the consumer are independent. In all such cases, the Discom shall only enter into agreement with the consumer.

c. **Capacity Limits and Interconnection Voltages**

For a gross-metered system there are capacity limits specified for different system sizes in kW (or MW) to be connected to the grid at appropriate voltages. These are typically in line with the state grid/supply code.

For a net-metered system, the interconnection is always with consumers' internal grid at LT level irrespective of system capacity.

d. **Procedure and Process**

Regulations do not need to contain a detailed process flow pertaining to application and approval process. This is typically in the purview of the implementing Discom, and the same should be duly indicated in the regulation. In addition, the regulation can also specify time limits for specific steps of the process to ensure timely and efficient implementation by Discoms and avoid grievances from consumers.

The following clauses are normally used to highlight the procedures in the regulation:

The distribution licensee shall allow connectivity to the solar PV rooftop system on first come first serve basis subject to operational constraints.

Provided that the available capacity at a particular distribution transformer, to be allowed for connectivity under these regulations, shall not be less than the limits as specified by the commission from time to time.

The distribution licensee shall provide information regarding distribution transformer-level capacity available for connecting solar PV rooftop system under net/gross-metering arrangement within one month from the date of notification of these regulations on its website and shall update the same within seven working days of the subsequent financial year under intimation to the commission.

The capacity of renewable energy system to be installed at any premises shall be subject to:

1. *The feasibility of interconnection with the grid.*
2. *The available capacity of the service line connection of the consumers of the premises.*
3. *The sanctioned load of the consumer of the premises.*

The distribution licensee shall formulate a detailed, transparent online procedure for application, registration, and grant of approvals for consumers who wish to install solar PV rooftop systems in the area of the distribution licensee.

e. **Grid Connectivity, Standards, and Safety**

The regulation must point to the CEA 'Technical Standards for Connectivity of the Distributed Generation Resources' Standards 2013.

The regulation must also point to the CEA 'Measures Relating to Safety and Electricity Supply' Standards 2010.

Safe solar PV penetration levels must be mentioned on a distribution transformer-basis.

The following clauses are normally used in this regard:

The solar PV rooftop generator shall be responsible for safe operation, maintenance, and rectification of any defect of the PV system up to the point of tariff meter beyond which the responsibility of safe operation, maintenance, and rectification of any defect in the system including the gross/net meter shall rest with the distribution licensee.

The distribution licensee shall have the right to disconnect the solar rooftop PV system at any time in the event of possible threat/damage, from such renewable energy system to its distribution system to prevent an accident or damage. Subject to Regulation 4 (2) above, the distribution licensee may call upon the renewable energy generator to rectify the defect within a reasonable time.

The distribution licensee shall ensure that the cumulative installed capacity on any distribution transformer shall not exceed 100% of the transformer rating in kVA or MVA. Once this penetration limit has been reached, the distribution licensee must carry out a detailed load flow study before granting any further connection approvals.

f. **Metering**

The regulations must point to the CEA 'Installation and Operation of Meters' 2010.

The metering arrangement and jurisdiction (who shall procure and own the meter, etc.) must be clearly laid out in the regulations.

The type of meter should be specified (bi-directional meter, accuracy class, etc.) and cost for the meter should be apportioned to the relevant stakeholder (consumer or Discom).

The responsibility for charges for installation and testing of the meter should be also clearly apportioned.

The regulation may also specify different accuracy class of meters depending on the type of consumer (residential, commercial, and industrial). Time-of-Day (TOD-based meters are also usually specified for industrial consumers. It is recommended to maintain the same accuracy class of the gross/net meter as the consumer's earlier conventional meter.

g. **Energy Accounting, Billing, and Banking**

Energy accounting, billing, and banking that are essential for settlement of excess energy are also considered in the regulation. These may include as per the need as follows:

- Differentiation between residential, industrial, commercial, and other types of consumers
- Differentiation between open access consumers, captive, self-owned systems, and third party-owned systems
- What is the settlement period (one month/billing cycle/one year/15 minute
- What is the financial incentive in case the consumer is net positive in export of energy generated by the solar system in the specified settlement period
- What are the charges for banking excess energy on the grid
- What are the withdrawal charges (in INR/kWh) during peak load times
- Applicability of open access charges (if needed) for third-party-owned rooftop systems

h. **RPO**

One of the main drivers of any solar programmes, whether on rooftop or on the ground, is the RPO. In addition to the (i) Discom, this RPO is also applicable to (ii) consumers with large captive power plants, usually greater than 1 MW, and (iii) open access consumers with large contract demands, usually greater than 1 MW. These 'obligated entities' are defined by the SERC from time to time.

Solar PV rooftop plants directly cater to the RPO. However, many consumers may not be obligated entities, and in this case, the Discoms may be encouraged by accounting all the generated solar energy towards the Discom's RPO.

The following clause for RPO may be found in the regulation:

Distribution licensee shall claim RPO if (i) consumed solar energy is not credited towards the RPO of the consumer or any other third party and (ii) no REC is claimed for the generated solar energy.

3.6 BUSINESS MODELS FOR ROOFTOP SOLAR DEPLOYMENT

3.6.1 Introduction and Significance of Business Models

A business model is a plan implemented by a company or an organisation to deliver a value-based proposition (product or a service or a combination of the two) to a customer with the objective of earning revenues and profit. The business model formulates and communicates the logic behind the value created and delivered to the consumers. In essence, a business model is a conceptual, rather than financial, model of a business.

Design of appropriate business models assumes a greater significance in the case of solar PV rooftop market due to their relatively high cost of energy generation/high upfront investments coupled with distributed implementation and generation. Hence, appropriate design and packaging of a solar rooftop deployment programme in terms of a viable business models is key to its success and should be the basis of any policy or regulation formulation.

3.6.2 Metering Arrangement

The Net-Metering Concept: Net metering is the concept which enables the eligible consumer, generates electricity using solar rooftop or other forms of renewable energy installed at its own premises for self-consumption, and banks the excess electricity with the grid.

Excess energy that may be in the form of energy credits can be used to offset the electricity consumed by the consumer in next month energy consumption from the grid. In some cases (like Karnataka), the excess energy is bought by the utility at a regulator defined feed-in tariff.

This whole process is made possible through the use of a bi-directional meter which records the net import and export of energy. On the basis of net consumption shown in the meter, the consumer pays the bill as per the predetermined tariff.

The gross and net-metering arrangements are illustrated below.

Gross-metering arrangement: Energy generated from rooftop system is exported to the grid. Energy for self-consumption is imported from the grid. The user gets two line items on their bills: one of energy consumption and another for exported energy.

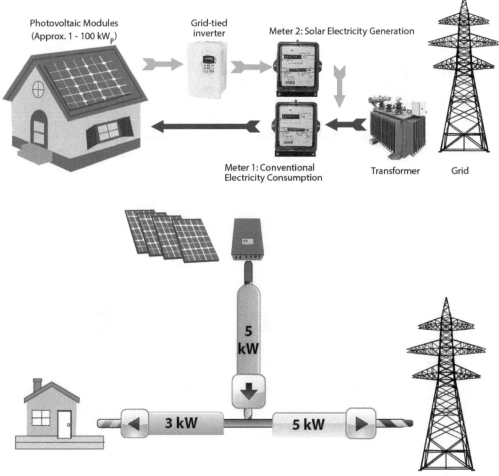

Utility pays for 5 kW Gross Generation by the consumer

Figure 3.29: *Gross-metering arrangement*

Net-Metering Arrangement: Energy generated by the rooftop system is first used internally and excess electricity is exported to the grid through a bi-directional meter. The bill received by consumer is net of electricity imported from the grid minus electricity exported to the grid.

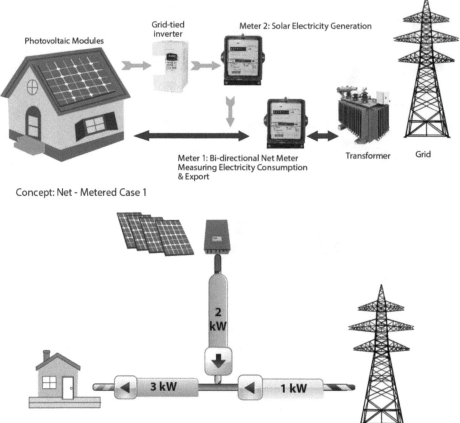

Concept: Net - Metered Case 1

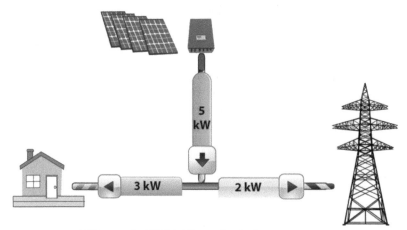

Utility bills for 1 kW Net Consumption to the consumer

Concept: Net - Metered Case 2

Utility pays for 2 kW Net Generation by the consumer

Figure 3.30: Net-metering arrangement

The scale of integration of new electricity generation sources in national grid network will present both challenges and opportunities for utilities in India and also lead to the creation of new business models as the worldwide trends suggest. An overview of the potential business models will be provided in the later sections.

3.6.3 Categorisation of Solar PV Rooftop Business Models:

The design of solar rooftop business models across the globe has been undertaken on the basis of two key parameters – ownership structure and revenue structure. These two are backed up in some cases by a third variable which we can call other drivers and which basically plug the gap between the cost of ownership and the revenues by providing appropriate fiscal incentives. Figure 3.31 highlights the key variables which have been used across the globe for the design of solar rooftop business models.

The variable which has shaped the solar rooftop business model space has been the ownership structure. Ownership structure has defined the manner in which the rooftop systems are owned but have also provided the framework within which these structures can be made economically and financially viable in combination with a host of other parameters.

Once the ownership structure has been decided, it needs to be coupled with a revenue model. Revenue models depend on the manner in which the energy is generated and used/sold. This is followed by other incentives which may be needed for ensuring the financial viability of the business model.

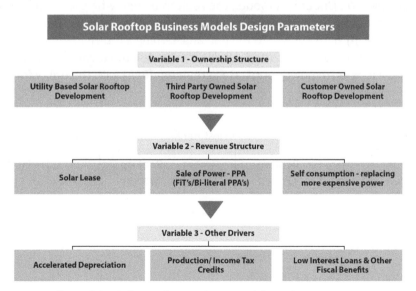

Figure 3.31: Solar Rooftop Business Models Design Parameters

3.6.3.1 Self-Owned Business Models

Self-owned business models, as the name suggests, promote investment in solar rooftop systems by the end users of solar energy themselves. Self-owned business models for grid-connected solar PV rooftop deployment have developed through the following two routes:

1. Gross metered
2. Net metered

Systems developed under self-owned business models either generate electricity for onsite consumption or for export to the grid. For most of the self-owned business models, the rooftop owner invests the equity component of the rooftop system while the debt component is usually financed through a FI such as a commercial bank.

a. **Gross Feed**

Gross feed-based solar rooftop systems consist of grid-connected solar rooftop systems which feed all the energy generated to the grid. In lieu of the energy fed to the grid, they are paid feed-in-tariff (FIT).

(i) Design: Self-owned gross feed rooftop installations, first adopted on a large-scale by Germany, are amongst the most popular across the globe. Under this model, a regulator approved FIT is used for the procurement of all the power generated by a solar PV rooftop system. The basic assumption for this market to work is that the FIT provides a minimum rate of return on the investment to the investor. Under this model, the rooftop owner, who is also the consumer to the utility, installs a solar rooftop system with the intention of exporting (feeding in) all the power to the grid and earning a return in the form of FIT for each unit of power exported. Under this model, the owner/consumer enters into a long-term power purchase agreement (PPA) with the utility for the sale of power.

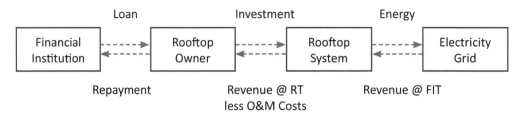

Figure 3.32: Design of Gross Feed Business Model

(ii) <u>Application</u>: The key markets to adopt gross FITs for solar rooftop systems have been Germany, Italy, France, other European Union nations, Japan, and the Gandhinagar Solar Rooftop Pilot project in India.

(iii) <u>Ownership</u>: All systems under this model are owned by the consumers themselves. In mature solar rooftop markets, most of these systems are eligible for project financing with no collateral from the rooftop owner.

(iv) <u>Advantages</u>: The gross FIT model offers three main advantages:

- The biggest advantage (for a gross or net-metered grid-connected solar rooftop system) is that these systems do not need to be coupled with stand-alone storage devices which bring down the cost of energy generated.

- The gross FIT model allows entry of a number of new investors (consumers), resulting in the enhancement of the investment base. The gross FIT mechanism also safeguards utilities against migration of high-paying consumers out of the utility ecosystem and safeguards the long-term viability of the grid.

- Under the gross FIT mechanism, the utility procures the solar rooftop power (which is at a higher cost than conventional power), the cost of which is passed on a part of the Annual Revenue Requirements (ARR) of the utility and socialised across all consumers being serviced by the utility.

- Gross FIT also allows all consumer categories, regardless of their consumer tariffs to participate in the solar rooftop programme and develop optimally sized solar rooftop installations and earn a minimum rate of return on the investment made by them.

(v) <u>Disadvantages:</u> The FIT is usually higher than the average power purchase cost and creates an apparent short-term cash flow burden on the utility's balance sheet. The higher cost of procurement also leads to increases in consumer tariffs.

b. **Net Metering**

(i) <u>Design</u>: Self-owned net-metering rooftop installations are amongst the most popular business model followed in the United States (U.S.). Under the net-metered business model, solar energy is first consumed by the consumer for meeting internal/captive loads and rest is banked with the utility and is subsequently netted of against imports from the grid. Most utilities and regulators aim to regulate the size of the systems in such a way that the generation of the system is lower than the annual energy demand of the rooftop owner's energy requirements.

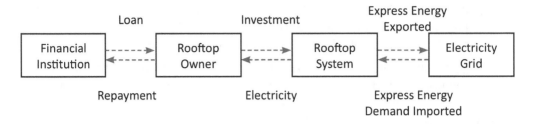

Figure 3.33: Design of net-metered business model

(ii) Application: The value proposition from this model comes from the difference between the consumer tariffs and the cost of solar generation from these installations. If the consumer tariffs are higher than the cost of solar rooftop installations, then the net-metering mechanism and the associated business models become quite attractive. In case tariffs are lower than the cost of generation, then installations do not take place or have to be incentivised through fiscal incentives.

(iii) Ownership: All the systems under this model are owned by the rooftop owners/consumers themselves.

(iv) Revenue Stream and Benefits: There are effectively two revenue streams in this model. The first is based on the savings due to the avoided cost of power purchase from the grid. The second is the sale of surplus power generated (at a rate determined by the regulator) over and above the consumer's own consumption within a settlement period.

(v) Advantages: The net-metered business model offers the following two main advantages:

- The net-metered model does not depend on any high FIT (which is usually higher than the average power purchase cost for the utility) and thus does not cause any significant outflow of funds from the utility.

- The net-metered model allows only those consumers to install solar rooftop who can afford to pay for solar and discourages socialisation of higher solar tariffs thus bringing down the impact of high solar costs across the whole cross-section of consumers.

(vi) Disadvantages: The net-metering framework works on the premise that solar power replaces the more expensive grid power. Therefore, the net-metering concept works only for consumers with high grid tariffs. This framework has a severe limitation in a market like India where the cost of power for a large majority of the consumers is below the cost of solar power like domestic or institutional consumers. The net-metering framework also reduces the net quantum of power sold by utilities to consumers, especially for high-tariff -paying consumers.

3.6.3.2 Third-Party-Owned Business Models

Under the-third party-owned model, a third party (separate from the consumer [rooftop owner] and the utility) is the owner of the rooftop systems. This third party may lease the rooftop from the rooftop owner and then generate power which may be sold to the utility or to the rooftop owner through a PPA, or the third party may lease the system to the rooftop owner who may utilise power from the system.

Third-party-owned models are emerging as a significant market force in the solar rooftop segment due to certain inherent capabilities that they bring to the business, such as access to low-cost financing; greater ability to take on, understand, and mitigate technical risks; aggregate projects and bring in economics of scale; effectively avail tax benefits; and the ability to make use of all government incentives.

Third-party-owned rooftop systems have been developed through the following two main routes:

- Solar Leasing
- Solar PPA

a. **Solar Leasing**

(i) Design: Solar leases were initially introduced in the U.S. market for financing residential PV systems where rooftop owners leased solar PV rooftop systems from large profit-making investors who acted as lessors. The lease agreement stipulates that the rooftop owner would make a monthly lease payment to the lessor over a specified period of time while enjoying the benefit of the electricity generated from the system while benefiting from lower utility bills.

> **Case Study of Solar Leasing: SolarCity**
>
> SolarCity is one of the largest solar lease companies operating in the United States SolarCity provides residential solar leases, which are financed by FIs and equity investors who claim income tax credit and depreciation benefits. Solar City offers its customers a variety of lease structures including zero down-payment options. The lease payments cover the cost of the system and the cost of monitoring, maintenance, and repair including inverter replacement, if necessary. SolarCity also guarantees a minimum level of electricity output from the rooftop PV system.

(ii) Ownership: The ownership of the solar rooftop systems continued to lie with the lessor. The leases are usually drafted for a fixed period of time and at the end of the lease period, the rooftop owner has the option to (a) purchase the PV system, (b) extend the lease agreement, or (c) remove the system from the roof. The lease arrangement provides an option for home owners to benefit from solar power while not making an upfront payment for the same.

(iii) <u>Revenue Streams and Benefits</u>: The third-party investor earns steady cash flows in the form of lease rental payments while also benefiting from tax credits and depreciation benefits available to investors of solar rooftop equipment. The tax benefits help shore up project internal rate of return (IRR) which in turn brings down the cost of leasing the systems to home owners.

(iv) <u>Advantages</u>: The key advantage of solar leasing solutions lies in (a) the rooftop owner is not required to make an upfront investment in solar rooftop systems but still benefits from use of these systems and (b) the use of tax benefits makes these systems cheaper to the rooftop owner.

> **Case Study of Net Metering: SunEdison**
>
> A well-known example for net metering on individual rooftops with third-party-owned systems with grid feed is the SunEdison LLC's agreement to supply power to Walmart Stores using the latter's rooftops at several locations. These PPAs were sold to investors who then became the owners of the installation and could claim tax credits and rebates on the investments. In its current form, this model is restricted in its application by factors like rooftop space, need for peak-differentiation in retail tariffs (e.g., through time-of-use tariff schemes, etc.) for viability and certain minimum day-time demand exceeding solar generation.

(v) <u>Disadvantages</u>: The two main disadvantages to the solar leasing framework for India are:

- Leasing of capital equipment like solar rooftop systems attracts a service tax, which makes leasing uncompetitive over the life of the project.

- There may be no relation between the lease rental paid by the consumer to the lessor and the quantum of energy generated from the systems. There is a need to benchmark lease payments with a minimum performance in terms of generation from leased systems.

b. **Solar PPAs**

<u>Design</u>: Under the third-party ownership model third party developers invest in solar rooftop assets which can then be sold either to the building owner (also the utility consumer) or fed into the grid. Third-party development models work because they have the wherewithal to aggregate rooftops, structure large projects which bring economies of scale and also take maximum leverage of government incentives, driving down the cost of solar power.

<u>Application and Revenue Arrangements</u>: A number of commercial arrangements have come into the market where third-party developers sell the power either the rooftop owner or to the grid through a PPA. Some of these arrangements have been highlighted below:

(i) <u>Individual Rooftops With Third-Party-Owned Systems With Grid Feed</u>:

- <u>Gross Metering With Third-Party Ownership of Systems</u>: Under the gross-metering arrangement, the third-party developer leases a rooftop and pays a rooftop lease/rental for the rooftop space. The developer exports the solar energy generated from the rooftop installation to the utility at a predetermined FIT set by the regulator. The key challenge in this model lies in the availability of the rooftop for 25 years.

- <u>Net Metering with Third-Party Ownership of Systems</u>: Under this arrangement, the rooftop owner signs a PPA with the third-party developer (who is given the rooftop for the installation) and enters into a back-to-back net-metering arrangement with the utility. The rooftop owner buys electricity generated by the third-party developer at a fixed price under a long-term PPA. This model is quite prevalent in the United States; especially with large energy consumers such as retail chains or warehouses and logistics companies. This model has become quite successful in markets which have a high cost of electricity and time-of-day tariffs.

(ii) <u>Combined Rooftop Leased by Third-Party with Grid Feed (Gross Metering)</u>: Under this model, a project developer identifies and leases (through a lease agreement) a number of rooftops in an area and develops these together in the form of a single project. The project developer invests in equipment, sets up the project, and sells the energy generated to the utility. This model was followed for the pilot demonstration solar rooftop project under the Gandhinagar Solar Rooftop Programme, where all the energy generated by the systems is being fed into the grid and the rooftop owners are entitled to a generation-based lease rental.

Case Study: 5 MW Gandhinagar Solar Rooftop Programme

The 5 MW Gandhinagar Solar Rooftop Programme is a successful example of combined rooftop leased by a third-party with grid feed model via gross metering. This is among the first programmes to implement solar rooftop at a megawatt-scale in India, that too as a PPP.

Here, two solar project developers, Azure Power and SunEdison, were selected through a tariff-based reverse bidding and given a quota to install an aggregate of 2.5 MW of solar PV rooftop systems each. The two developers signed a PPA with Torrent Power Limited, the local distribution utility of Gandhinagar. Power generated by each solar rooftop system is fed into and accounted for using a dedicated feed-in meter. Rooftop lease agreements between the project developer and the rooftop owner, whether private residential or government, were designed.

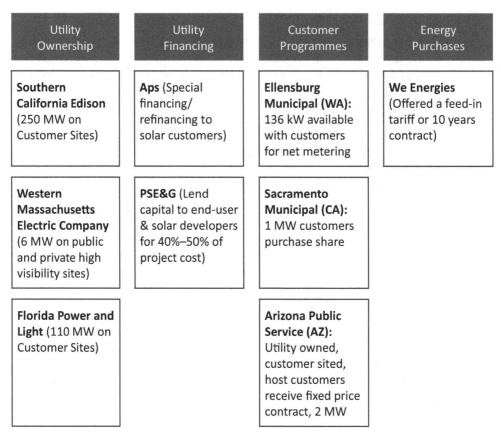

Utility Ownership	Utility Financing	Customer Programmes	Energy Purchases
Southern California Edison (250 MW on Customer Sites)	**Aps** (Special financing/ refinancing to solar customers)	**Ellensburg Municipal (WA):** 136 kW available with customers for net metering	**We Energies** (Offered a feed-in tariff or 10 years contract)
Western Massachusetts Electric Company (6 MW on public and private high visibility sites)	**PSE&G** (Lend capital to end-user & solar developers for 40%–50% of project cost)	**Sacramento Municipal (CA):** 1 MW customers purchase share	
Florida Power and Light (110 MW on Customer Sites)		**Arizona Public Service (AZ):** Utility owned, customer sited, host customers receive fixed price contract, 2 MW	

Figure 3.34: Utility-Based Business Model

3.6.3.3 Utility-Based Business Models

Utility involvement in the solar rooftop market was initially limited to being a facilitator through a broad framework for interconnection. Some utilities also provided solar PV systems and are associated with these systems, but this is limited to municipal utilities. However, a growing number of investor-owned utilities have recently taken up a more proactive stance in encouraging the development of solar rooftop projects due to (1) reduction in price of distributed clean energy technologies like solar PV; (2) advent of a number of investors and developers who can implement

> **Case Study of Customer-Sited PV: San Diego Gas and Electric**
>
> San Diego Gas & Electric (SDG&E), encourages development of solar rooftop installations, owned by SDG&E and installed on leased rooftops of customers. The rooftops are leased by SDG&E, generally for 10 years, with a possibility of two 5-year extensions. The rooftop systems are connected using a gross-metering format.

systems which partially or wholly replace the grid; (3) maturing technology and easy financing options that allow consumers to partially switch to these technologies, and (4) proactive policy and regulatory frameworks that allow these decentralised distributed technologies to come into play in the market.

Keeping the impact of these disruptive technologies like solar PV rooftop in mind, the utilities have also started working towards active participation in these emerging segments through the use of new and innovative utility-based solar PV rooftop business models which aim to capture value of these markets. The utilities involvement in the solar PV rooftop business model space has been limited to four broad areas which have been highlighted in Figure 3.34 with relevant examples:

a. **Utility Ownership**

Utilities are becoming more and more aggressive in owning rooftop systems as it allows them to claim tax credits and earn a healthy rate of return on the power generated from these installations while also ensuring that consumers with rooftops do not transit out of the utility's ecosystem.

A number of utilities ranging from San Diego Gas and Electric, Southern California Edison to Western Massachusetts Electric Company are aggressively developing rooftop installations on customer sites. The overriding reason behind the success of this model is the regulated rate of return that is available for these utilities for the capital investment in rooftop installations.

Case Study of Utility Financing: Powder River Energy Corporation

Onbill financing was offered by Powder River Energy Corporation, Wyoming, to its residential customers – they could take loans up to USD 2,500 at a 0 % rate of interest for up to 36 months. The Public Service Electric and Gas Company (PSE&G) of New Jersey also offers utility-based loans at 6.5 % for up to 10 years and covers around 40 to 60 % of the system cost.

b. **Utility Financing**

Another route in which utilities are encouraging the deployment of solar rooftop installations is by financing consumers. Utility and public financing programmes have been launched by a number of utilities and local governments across the United States to facilitate adoption of solar PV with two broad aims of (a) covering rooftop owners who do not have access to traditional financing options (self/third party) and (b) enhancing affordability of systems by reducing interest rates and upfront fees and relaxing lending guidelines. Two broad types of loans are available through utility-based financing:

(i)   Utility Loans: These are loans are targeted at utility customers and administered by the utility at the local, municipal, or the state level. These programmes are structured so as to be either cash-flow positive or neutral in order to make electricity savings equal to or greater than the cost of the loan. Utility loans are either linked to the consumer (bill financing) or linked to the property (meter-secured financing).

(ii) Revolving Loans: Revolving loans finance rooftop owners directly through public sources such as public benefit funds, environmental non-compliance penalties, bond sales, or tax revenues. Rooftop owners prefer these as they come at low interest rates and have relaxed lending guidelines and extended tenors. The Montana Alternative Energy Revolving Loan Program is one such example.

c. **Community-Shared or Customer Programmes**

Community share solar programmes provide energy consumers the option of utilising the benefits of solar generation (through proportional benefits via virtual net metering) without actually installing on-site solar PV or making high upfront payments required for such projects. These plants are usually set up by community-owned utilities or third parties in partnership with investor-owned utilities.

The community share programme allows utilities to develop larger programmes and projects while providing expanded options to more customers at lower costs. The broad outline of a community-shared solar project model is highlighted in Figure 3.35.

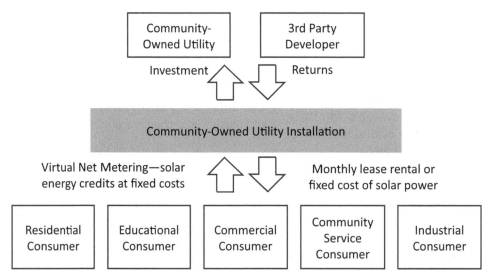

Figure 3.35: *Community-Shared Utility-Based Business Model*

The community members who sign up for these projects receive solar benefits without paying upfront capital cost, installation cost, or worry about the O&M.

d. **Energy Purchases**

A number of utilities are also entering the market with the objective of procuring energy directly from third party or rooftop owners by offering FITs which allow utilities to buy all the energy generated by the rooftop at a flat price under a long-term PPA, the cost of which is passed onto the consumers as part of its ARR while at the same time retaining the customers on whose rooftops these systems have been set up.

3.6.4 Key Challenges and Considerations

While the Indian solar PV rooftop market provides a number of opportunities for a host of developers/investors, the design and implementation of business models in India still remains a challenge, especially for third-party developers who want to bring in greater scale and efficiency into the rooftop development market.

Examples of Community Share Programmes Offered in the United States

1. **Tucson Electric Power's (TEP) Bright Tucson Community Solar Programme:** TEP, an investor-owned utility operating in Arizona, United States, launched a third- party developed community-based solar program for developing 1.6 MW of new solar capacity in three years. This programme allowed consumers to buy generating blocks of 150 kWh per month for a monthly fixed fee of USD 3. The investments for the solarinstallations were made by a third-party developer.

2. **Colorado Springs Utility's Community Solar Gardens Program:** In 2010, the Colorado Spring Utility offered its customers the chance to invest in community solar gardens. Under this scheme, the customers could lease panels from one of two community solar project developers, Sunshare or Clean Energy Collective, with a minimum solar garden interest of 0.4 kW. All customers who subscribed to the programme received a fixed credit of USD 0.09/kWh on their electric bill for their share of the power generated by the panels they had leased. The pilot run by the programme was for 2MW of installations.

3. **Sacramento Municipal Utility District (SMUD) Solar Shares Program:** SMUD's SolarShares Program provides an opportunity to customers who cannot or choose not to acquire PV systems on their own to purchase solar power directly from the installations under SMUD's SolarShares Program. The programme procures solar power from third-party developers or community-based solar installations and passes these onto the consumers. SMUD pays a fixed tariff for the power and then resells the solar power to participating customers who get credits for the solar power using a virtual net-metering scheme.

Two Examples of Energy Purchase by Utilities

1. **We Energies Feed-In Rate:** We Energies, a utility serving in Wisconsin and Michigan's Upper Peninsula offers a FIT similar to the solar FITs offered by European markets like Germany. The FIT offered by the utility is USD 0.225/ kWh for 100% of the solar power generated, with the customer getting a credit on its bill or a check

2. **Gainesville Regional Utilities' (GRU) FIT:** GRU, a municipal Utility in Florida, offers a FIT as an alternative to the rebate programme which allows it to not lose utility customers, spread rebates over longer periods, and the contracting is performance-based

Solar rooftop projects suffer from a number of commercial, policy and regulatory, technical, and financing challenges which need to be addressed as the market grows through a concerted effort from policy makers, regulators, financers, and above all the utilities.

Some of these challenges have been highlighted below:

a. Contract Sanctity: One of the major challenges that developers face in the Indian solar PV rooftop market are of long-term contract sanctity. Third-party developers have to enter long-term contracts with rooftop owners which are mostly backed up by Letters of Credit for one month's billing and with limited long-term payment security. For long-term sustainability and enhancing investor confidence in the market, there is a need to ensure that the contracts are honoured over the project and also ensure that appropriate legal options are available for both parties in case of premature termination of contract.

Contracts need to be easily enforceable, provide remedies for payment defaults, and buy out clauses/appropriate compensation framework in case of building redevelopment or relocation of projects.

b. Availability of Financing and Capacity of FIs to Evaluate Rooftop Projects: Access to project financing and consumer financing is one of the key requirements to scale up the solar rooftop sector. Banks and FIs are still in the process of putting in place consumer financing products (loans) and guidelines which allow access to debt for rooftop owners. In case of third-party developers, especially in the commercial and industrial space, banks and FIs still lack appropriate tools and expertise to evaluate these projects especially from a long-term risk perspective. As new business models come into the market, banks and FIs will also have to increase their capacity to analyse and finance these models.

c. <u>Solar Equipment Leasing</u>: One of the key fiscal incentives used to bring down the cost of solar in markets like the United States is depreciation or accelerated depreciation (AD) in the case of India. This benefit is not available to new special purpose vehicles (SPVs) but can be utilised through investors who can buy the equipment and then on-lease the same equipment to developers. The key challenge here is that service tax is levied on the leased equipment which erodes most of the benefits that investors may have attained from AD.

d. <u>Rooftop Leasing</u>: Access to rooftops for the life of the solar rooftop project remains another key challenge due to issues such as reconstruction of the building or expiry of the lease of tenets who have signed the PPA. Most private sector companies lease buildings (along with rooftops) for up to 10 years. Developing rooftop projects on buildings with leases up to 10 year becomes risky in case the next tenet or the building owner does not agree to extend either the lease or the PPA. Risk to rooftop projects also comes in when rooftop owners might want to construct more floors or reconstruct the whole building before the lease/PPA runs its natural life. Cases like these have come to light in New Delhi where institutions are not ready for solar rooftop despite a very competitive tariff and adequate space.

e. <u>Role of Utilities – Challenges and Facilitation Required</u>: One of the biggest challenges facing the solar rooftop sector is the limited capacity of the utilities in implementing solar PV rooftop projects. Interconnection processes are slowly being specified and, in some cases, are long and cumbersome often allowing only a few contractors/developers to participate.

There is a need to streamline the interconnection process, making them time bound and transparent with a focus on achieving the required performance standards and quality standards. One example is of the Bangalore Electric Supply Company Limited (BESCOM), which using an open sourcing framework specified the need to only adhere to national and international standards while deploying systems and interconnecting them to the grid.

f. <u>Match between Incentive Mechanisms and Needs of the Market</u>: The policy makers and regulators have chosen the net-metering framework for promoting solar PV rooftop development in India. While this framework has a number of advantages, it also suffers from the basic challenge of not allowing all consumer categories to develop solar rooftop projects. There is a need to evaluate a regulatory framework which targets rooftop utilisation and penetration of solar rooftop systems on a large number of consumers such as schools, hospitals, and storage facilities, which have huge rooftop space but do not have the financial justification for adopting net-metered solar rooftop business models.

UNIT 4 PRE-COMMISSIONING INSPECTION OF GRID-CONNECTED ROOFTOP SOLAR PV SYSTEM

It is essential that the engineer checks and confirms all the documents submitted by the consumer (and its installer) pertaining to component certifications, compliance with state regulation and/or Discom guidelines.

There are many component certifications specified in many state regulations or in Discom interconnection processes and these shall be conveyed to the consumer in approval letter. These are to be complied by the consumer and its installer for all grid-connect rooftop PV systems. Even in cases where these are not specifically mentioned in the documents, all parties have to comply with the minimum certifications as decided by the Ministry of New and Renewable Energy (MNRE). Modules and Inverters are the major components of these systems and standards and certifications for these are as tabulated:

Solar Modules	
IEC 61215/ IS 14286	Design Qualification and Type Approval for Crystalline Silicon Terrestrial PV Modules
IEC 61701	Salt Mist Corrosion Testing of PV Modules
IEC 61853- Part 1/ IS 16170: Part 1	PV module performance testing and energy rating: Irradiance and temperature performance measurements and power rating
IEC 62716	PV Modules – Ammonia (NH3) Corrosion Testing (As per the site condition like dairies, toilets)
IEC 61730-1,2	PV Module Safety Qualification – Part 1: Requirements for Construction, Part 2: Requirements for Testing

Solar Modules	
IEC 62804	PV modules - Test methods for the detection of potential-induced degradation. IEC TS 62804-1: Part 1: Crystalline silicon (mandatory for applications where the system voltage is > 600 VDC and advisory for installations where the system voltage is < 600 VDC)
IEC 62759-1	PV modules – Transportation testing, Part 1: Transportation and shipping of module package units

Solar PV Inverters	
IEC 62109-1, IEC 62109-2	Safety of power converters for use in PV power systems: Part 1: General requirements and safety of power converters for use in photovoltaic power systems Part 2: Particular requirements for inverters safety compliance (Protection degree IP 65 for outdoor mounting, IP 54 for indoor mounting)
IEC/IS 61683 (as applicable)	Photovoltaic Systems – Power conditioners: Procedure for Measuring Efficiency (10%, 25%, 50%, 75% & 90–100% Loading Conditions)
BS EN 50530 (as applicable)	Overall efficiency of grid-connected photovoltaic inverters: This European Standard provides a procedure for the measurement of the accuracy of the maximum power point tracking (MPPT) of inverters, which are used in grid-connected photovoltaic systems. In that case, the inverter energises a low voltage grid of stable AC voltage and constant frequency. Both the static and dynamic MPPT efficiencies are considered.
IEC 62116/ UL 1741/ IEEE 1547 (as applicable)	Utility-interconnected Photovoltaic Inverters - Test Procedure of Islanding Prevention Measures
IEC 60255-27	Measuring relays and protection equipment – Part 27: Product safety requirements
IEC 60068-2 (1, 2, 14, 27, 30 & 64)	Environmental Testing of PV System – Power Conditioners and Inverters a) IEC 60068-2-1: Environmental testing - Part 2-1: Tests -Test A: Cold b) IEC 60068-2-2: Environmental testing - Part 2-2: Tests - Test B: Dry heat

Solar PV Inverters	
IEC 60068-2 (1, 2, 14, 27, 30 & 64)	c) IEC 60068-2-14: Environmental testing - Part 2-14: Tests - Test N: Change of temperature d) IEC 60068-2-27: Environmental testing - Part 2-27: Tests - Test Ea and guidance: Shock e) IEC 60068-2-30: Environmental testing - Part 2-30: Tests - Test Db: Damp heat, cyclic (12 h + 12 h cycle) f) IEC 60068-2-64: Environmental testing - Part 2-64: Tests - Test Fh: Vibration, broadband random, and guidance
IEC 61000 – 2,3,5 (as applicable)	Electromagnetic Interference (EMI) and Electromagnetic Compatibility (EMC) testing of PV Inverters

Similarly, other important components of any rooftop grid-connected PV systems also have standards that are to be followed by the consumer. These are as tabulated:

Earthing /Lightning	
IEC 62561 Series (Chemical Earthing)	IEC 62561-1 Lightning protection system components (LPSC) - Part 1: Requirements for connection components IEC 62561-2 Lightning protection system components (LPSC) - Part 2: Requirements for conductors and earth electrodes IEC 62561-7 Lightning protection system components (LPSC) - Part 7: Requirements for earthing enhancing compounds
Junction Boxes	
IEC 60529	Junction boxes and solar panel terminal boxes shall be of the thermo-plastic type with IP 65 protection for outdoor uses, and IP 54 protection for indoor use
Energy Meter	
IS 16444 or as specified by the Discoms	A.C. Static direct connected watt-hour Smart Meter Class 1 and 2 — Specification (with Import & Export/Net energy measurements)
Solar PV Roof Mounting Structure	
IS 2062/IS 4759	Material for the structure mounting

Important Notes:

> ➢ The state net-metering (or gross-metering) regulation has topmost priority and authority over other available national-or-state-level regulations and policy documents. In case of conflict between the documents, SERC regulation must be followed.

> ➢ It is also possible that all state regulations and policies may not specify standards and certification requirements for rooftop grid connect PV systems. In such cases, these standards are to be considered as guidelines and must be mandated only if stated clearly in interconnection process or approval letter.

> ➢ Else the Discom engineer may limit his document verification only to the standards related to solar module and inverter. These must always be followed.

> ➢ Additionally, in such cases, protection features of the systems must be checked and approved because these can have direct impact on health and quality of the grid to which these systems are connected.

Important Signages related to grid-connected rooftop PV systems

For safety of maintenance personnel, inspectors, Discom personnel, and emergency aid services it is essential to indicate markings of SPVRT installation. IEC 62548 indicates the following signs using local language or using appropriate local warning symbols

- Sign at the origin of the electrical installation
- Sign at the metering position, if remote from the origin
- Sign at the consumer unit or distribution board to which the supply from the inverter is connected
- Sign at all points of isolation of all sources of supply
- Sign at first upstream grid distribution point like pole, distribution transformer, bays in a substation, and so on

Some sample signs that are necessary to be put up are:

The engineer should understand that following regulations cover all technical aspects of grid connectivity of distributed generating sources.

- Central Electricity Authority (Technical Standards for Connectivity of the Distributed Generation Resources) Regulations, 2013
- IEEE 1547 – Standard for Interconnecting Distributed Resources with Electric Power System, 2003
- Central Electricity Authority (Measures Relating to Safety and Electricity Supply) Regulations, 2010

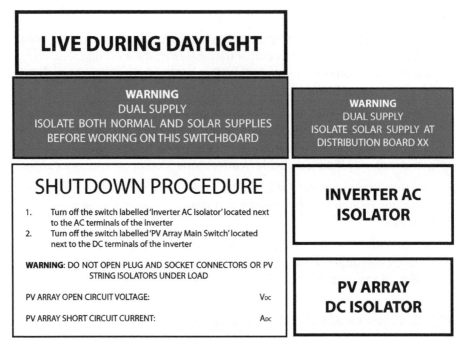

Figure 4.1: Sample Signs at the site

- Central Electricity Authority (Technical Standards for Connectivity to the grid) Regulations, 2007
- IEEE 519: Recommended Practice and Requirements for Harmonic Control in Electric Power Systems, 2014

These are accepted and adopted by all state electricity regulatory commissions and hence are binding for grid engineers in providing grid connection to rooftop PV systems.

There are important interconnection parameters as listed below that are monitored by grid-connected inverter/s in each system. Each inverter individually synchronises with the grid and therefore all inverters must comply with certificates mentioned here so that the interconnection parameters are also complied with.

Following are the main parameters:

General Requirements (IEEE 1547 / CEA Regulation 2013)

- **Voltage Regulations**
    - The distributed resources shall not actively regulate the voltage at the point of interconnection

- **Synchronisation**
  - o The distributed resource synchronised with electric system shall not cause a voltage fluctuation at the point of interconnection greater than ±5%
- **Monitoring Provision**
  - o Each distributed resource of 250 kVA or more at a single point of interconnection shall have provisions for monitoring its connections status, real-power output, reactive power output, and voltages at the point of interconnection

Parameters	Reference	Requirement
Overall Grid Standards	Central Electricity Authority (Grid Standard) Regulations, 2010	Compliance
Equipment	BIS / IEEE / IEC	Compliance
Meters	Central Electricity Authority (Installation and Operation of Meters) Regulation, 2006, Amendments thereof, OERC Generic Tariff Order, 2013	Compliance
Safety and Supply	Central Electricity Authority (Measures of Safety and Electricity Supply) Regulation, 2010	Compliance
Harmonic Current	IEEE 519 and CEA (Technical Standards for Connectivity of the Distributed Generation Resources) Regulations, 2013	shall not exceed the limits specified in IEEE 519
Parameters	Reference	Limits
Synchronisation	IEEE 519 and CEA (Technical Standards for Connectivity of the Distributed Generation Resources) Regulations, 2013	Every time the generating station must be synchronised to the grid. Voltage fluctuation < ± 5% at point of inter connection
Voltage	IEEE 519 and CEA (Technical Standards for Connectivity of the Distributed Generation Resources) Regulations, 2013	Voltage-operating window should be under operating range of 80% −110% of the nominal connected voltage. Beyond a clearing time of 2 second, the PV system must isolate itself from the grid

Parameters	Reference	Requirement
Flicker	IEEE 519 and CEA (Technical Standards for Connectivity of the Distributed Generation Resources) Regulations 2013	Should not cause voltage flicker in excess of the limits stated in IEC 61000
Frequency	IEEE 519 and CEA (Technical Standards for Connectivity of the Distributed Generation Resources) Regulations, 2013	When the distribution system frequency deviates outside the specified conditions (50.5 Hz on upper side and 47.5 Hz on lower side), there should be over and under frequency trip functions with a clearing time of 0.2 seconds.
DC injection	IEEE 519 and CEA (Technical Standards for Connectivity of the Distributed Generation Resources) Regulations, 2013	Should not inject DC power more than 0.5% of full rated output at the interconnection point
Power Factor	IEEE 519 and CEA (Technical Standards for Connectivity of the Distributed Generation Resources) Regulations, 2013	While the output of the inverter is greater than 50%, a lagging power factor of greater than 0.9 should operate
Islanding and Disconnection	IEEE 519 and CEA (Technical Standards for Connectivity of the Distributed Generation Resources) Regulations, 2013	In the event of fault, voltage, or frequency variations, the PV system must island/disconnect itself within IEC standard on stipulated period
Overload and Overheat	IEEE 519 and CEA (Technical Standards for Connectivity of the Distributed Generation Resources) Regulations 2013	The inverter should have the facility to automatically switch off in case of overload or overheating and should restart when normal conditions are restored
Paralleling Device	IEEE 519 and CEA (Technical Standards for Connectivity of the Distributed Generation Resources) Regulations, 2013	Paralleling device of PV system shall be capable of withstanding 220% of the normal voltage at the interconnection point

Notes:

- As it is, the inverter that takes care of grid synchronisation and all related safety and other parameters, it is important to check that inverters are certified for functioning as grid-connect inverters and carry all required certifications.
- In the process of pre-commissioning inspection, the inverter certificate validation will confirm to all the above parameters of grid connectivity.

There are other aspects of installed system that need to be checked as pre-commissioning test. All the design and safety aspects are detailed in earlier sections of this guide, which must be studied and understood by the Discom grid engineers. The checklist for the same is given below.

Sr. No.	Item type	Remarks
1	Capacity of installed system	Please check AC capacity of the system that is equal to inverter capacity. The solar array capacity can be same or higher as permitted by the selected inverter.
2	Installation Layout – is it as per drawing?	This will confirm the locations of all protective features of the system and it also rechecks the capacity installed as per approved capacity.
3	Inverter IS/IEC standards qualified	As stated in previous table; it is important to check applicability of each certificate to actual model as used in the system and the validity of all such certificates
4	PV panel IS/IEC standards qualified	As stated in above table; it is important to check applicability of each certificate to actual model as used in the system and the validity of all such certificates
5	PV isolators/PV cables IS/IEC standards qualified	As stated in previous table; it is important to check applicability of each certificate to actual model as used in the system and the validity of all such certificates
6	String level and cumulative protections are provided for over-current and surges	OCP needs to be checked for its specifications, locations, and standards. OCP needs to be installed for each string and also for each combination of strings. The capacity of OCP fuse needs to be 1.56 times the short circuit current of the string or string combination, depending on where this OCP is located.
7	Surge protections are provided at all specific locations in the system.	SPD need to be provided on DC side in AJB / SCB / DCDB and on AC side in ACDB/Grid interface panel, if not in-built in inverter on either side. The type of SPD to be used depends on inverter and also on distance between solar array and lightning arrestor. Normally, these will be Type 2 SPDs.

Sr. No.	Item type	Remarks
8	DC side isolator switch between solar array and inverter	It is essential to have facility to disconnect DC side from the inverter in case inverter needs to be worked upon during day time. Solar array would be energised and only by operating disconnect switch one can safely work on inverter.
9	AC disconnect manual switch provided with locking arrangement	AC side isolator needs to be installed before interconnection point and access to this disconnector/switch should be free to Discom personnel. It should be possible to easily isolate complete system from the grid in case of grid line maintenance or fault or possible accident to the grid.
10	Meters approved by concerned authority	In case the consumer has procured net meter and solar (generation) meter, it is mandatory that these meters are tested and certified by the Discom meter testing laboratory. It should be checked whether installed meters are certified and sealed.
11	Earthing protections for DC, AC, and lightning arrestors	It is not sufficient that earthing has been provided for DC, AC, and for lightning arrestor. It is necessary to check that AC and DC side earthing pits are connected underground to achieve equi-bonding. This will eliminate chances of high potential difference between different system components thereby causing sparking, burning, and accidents. Also, the earth pits must be checked to see that the pit soil really has low resistance. Ideally, the soil resistivity should not be higher than 2 ohms while preparing the pits, and this should be tested and confirmed. During their lifetime, the resistivity must be maintained below 5 ohms as per the standard. Some regulations and interconnection processes require redundancy in earthing, and in such cases, it must be checked whether this is provided by having two earth pits for each earthing points.
12	Lightning protection systems	LPS consist of lightning arrestor and its earthing that must be separated from other earthings in the systems. It should be checked that the components used meet the required standards and that the full array and other components are covered within the effective protection area of the arrestors.

Sr. No.	Item type	Remarks
13	Signages	It is equally important to have warning and specification signs at different components and stages of the grid-connected system. The signs must be as per net-metering regulation of the state. In case there is no specific mention of signages in the regulation, it is essential that Discom and installers/consumers follow the industry standard practice. These signages are crucial for grid engineers as well as any person who may be called in future to work on repairs or maintenance of the system and of the grid.

Note: Sample IEC Certificates of Module and Inverter are attached in Annexure 1.

UNIT 5 POST-COMMISSIONING INSPECTION OF GRID-CONNECTED ROOFTOP SOLAR PV SYSTEM

Once the above-mentioned component and other checks are performed, it is time to carry out the functionality tests post-commissioning of the system. As stated earlier, this stage may come after Electrical Inspectorate (EI/CEI) has approved the system or without such approval depending on the capacity of the system and state-level decision of requirement of EI approval for certain capacities of rooftop PV systems.

This stage will cover the following:

➤ Install the net meter, already tested and certified by Discom meter-testing laboratory, by replacing existing consumer meter with suitable model.

➤ This may include work of installing new CTs/PTs depending on the model of net meter and that of existing consumer meter. This may require separate enclosures and bus bar that must have been already installed by the consumer/installer.

➤ Switching ON the system connection to the grid.

The grid engineer must remember and confirm the following while conducting the commissioning tests:

• All commissioning tests shall be performed based on written test procedures of equipment manufacturer/system integrator

• A visual inspection shall be made to ensure that the system earthing is adequate as per standard/regulation

• A visual inspection shall be made to confirm the presence of the isolation device as required by standard/regulation

- The following initial commissioning tests shall be performed on the installed solar PV rooftop system and grid-connected inverter prior to the initial parallel connection to the grid:
  - Operability test on the isolation device
  - Unintentional islanding functionality
  - Cease to energise functionality

On completing the pre-test work and confirming the above conditions, the following needs to be tested:

Sr. No.	Item type	Remarks
1	Check whether solar power generation stops automatically when Discom supply shuts off	This is an important test and is known as Anti-Islanding conformity test. The detailed procedure for this test is provided later in this section. (Sample Image can be seen in Figure 5.1)
2	Confirm bi-directional flow recorded on net meter	The engineer needs to adjust loads in the consumer premises to test bi-directional recording of power by the net meter. • Once consumer load is lower than the generation, the meter should record power flow to the grid. • Once consumer load is more than the generation, the meter should record power flow from the grid.
3	Check 'Consumption (Import) only' mode operation	For most of the period of a 24 hours day, the meter shall operate in 'import only' mode when PV system is not generating. This recording also should be tested by shutting down the PV system.
4	Check operation of Solar (Generation) meter	In most states, solar 'generation' meter is also mandatory to be installed by consumer or by Discom. Wherever applicable, this meter also should be tested for recording power flow, following a similar method as stated above (A sample image of Solar Generation Meter is shown below.

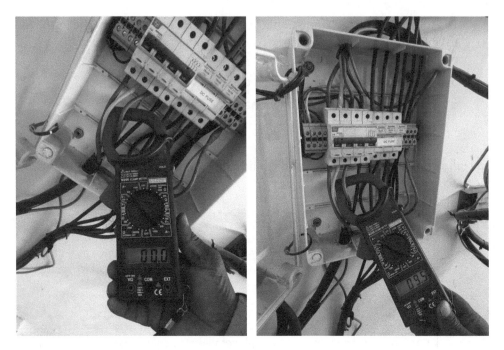

Figure 5.1: Anti-Islanding Test of Inverter

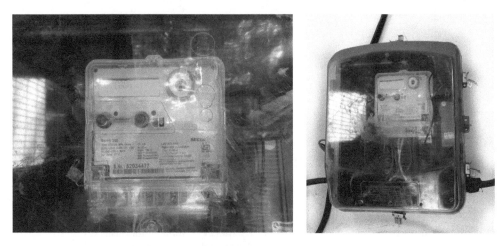

Figure 5.2: Solar (Generation) Meter

Detailed procedure of performing 'anti-islanding' test:

Requirements:

- ➢ This test must be performed during noon on a bright sunny day
- v   The PV system generation must be minimum 20% of the rated output power of inverter or solar array, whichever is less

> ➢ This test must be performed individually on all available inverters in the system

There are two parts of this test. First test is to check when the inverter shuts off when the grid is OFF. Second test is to check when the inverter wakes up when the grid is ON.

Test 1: Inverter must cease supplying power within 0.2 seconds of loss of mains

Step 1:
Keep DC supply from the SPVRT array connected to the inverter

Step 2:
Place the voltage probe in the inverter side of the AC main switch

Step 3:
Turn OFF the main switch through which inverter is connected to grid

Step 4:
Measure time taken for inverter to cease attempting to export power with a timing device and record

Test 2: Inverter must not resume supplying power until mains have been present for more than 60 seconds

Step 1:
Keep DC supply from the SPVRT array connected to the inverter

Step 2:
Place the current probe in the inverter side of the AC main switch

Step 3:
Turn ON the main switch through which inverter is connected to grid

Step 4:
Measure time taken for inverter to re-energise and start exporting power with a timing device and record

This date is recorded as system synchronisation date from which the metering and billing should start recording the net-metered system.

UNIT 6 MAINTAIN PERSONAL HEALTH AND SAFETY AT PROJECT SITE

6.1 ESTABLISH AND FOLLOW SAFE WORK PROCEDURE

6.1.1 Basics of Work Safety During Solar PV Installations

A solar PV system is to be installed as per the safety standards, installation, and building codes applicable for that particular location. The installer must be well versed with the organisation's safety policy. These considerations include:

- Maintaining a safe work area
- Safe methods of using tools and equipment
- Safety of personnel, which includes knowledge of personal protective equipment (PPE)
- Awareness of electrical and non-electrical hazards and how to mitigate them

General Requirements of a safe work area

- The work site must not be cluttered. This increases the chances of tripping or falling, especially on sloped roofs/surfaces.
- The access and exit arrangements should be safe and clear of obstacles.
- Communication arrangements should be adequate.
- Lighting should be adequate to allow safe working. Lighting in a work area may be a temporary arrangement cabled into the area requiring additional protection against possible trip-induced falls. Whenever possible, natural lighting should be provided in the work area during inspection.
- The tools must not be left unsecured as they may cause obstructions or fall off the roof, injuring someone below.
- The rooftop is an outdoor location. In case the location is extremely hot because of the sun, adequate precaution should be taken to avoid sun burns,

exhaustion, and de-hydration by use of sunscreen, wear light-coloured clothes, and keep drinking lots of water.

- Maintain a First Aid Kit to mitigate accidents involving personnel or any other person who may be in the vicinity (say, the customer)

m

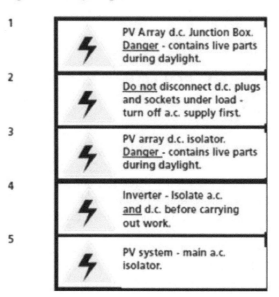

Figure 6.1: *Example of danger signs on components*

6.1.2 First Aid Kit

In spite of all precautions/preventions, there are chances of hazards. Therefore, a PV installer should always be aware of the mitigation measures to be taken if some mishap occurs. Following are the important mitigation measures that should be taken after the occurrence of hazards:

- Maintain composure and separate victim from cause of hazard
- Wash the injuries/wounds with clean water
- Apply First Aid on burns/injuries/wounds
- Apply Sodium Hydroxide (NaOH) solution when burns are due to battery's acid
- Take the injured person to a medical room or a safe place
- Look for the supervisor/ team leader
- Call the doctor/ medical officer

S. No.	Type of Hazard	Why It Happens
1	Loss of eye/vision	Using striking tools without eye protection.
2	Puncture wounds	Using a screwdriver with a loose handle which causes the hand to slip.
3	Severed fingers, tendons, and arteries	Using the wrong hammer for the job and smashing a finger.
4	Broken bones	Using a small wrench for a big job and bruising a knuckle.

Table 6.1: Common Tool Hazards

6.2 USE AND MAINTAIN PERSONAL PROTECTIVE EQUIPMENT (PPE)

6.2.1 Importance of PPE

Workplace hazards that could cause injury include the following:

- Intense heat
- Impacts from tools, machinery, and materials
- Cuts
- Hazardous chemicals

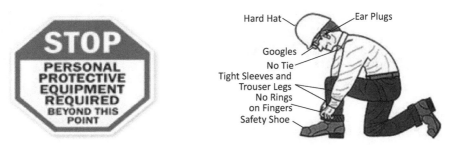

Figure 6.2: General PPE Guidelines

Ensure:

- Protect against specific hazard(s) encountered by employees
- Wear comfortable clothes
- Must not restrict vision or movement
- Durable and easy to clean and disinfect
- Must not interfere with the function of other required PPE

6.2.2 Eye/Ear Protection

For employees who use spectacles, they must use

- Goggles that can fit comfortably over corrective eyeglasses without disturbing their alignment
- Goggles that incorporate corrective lenses mounted behind protective lenses

Safety Glasses/Goggles   Ear Plugs

6.2.3 Head Protection

Protection from:

- Objects may fall from above and strike them on the head
- They may bump their heads against fixed objects, such as exposed pipes or beams
- They work near exposed electrical conductors

Head Protection Criteria:

In general, protective helmets, or hard hats, should:

- Resist penetration by objects
- Absorb the shock of a blow
- Be water resistant and slow burning

6.2.4 Foot and Leg Protection

Some of the potential hazards that would require foot and leg protection include:

- Heavy objects such as barrels or tools that might roll onto or fall on employees' feet
- Sharp objects such as nails or spikes that might pierce the soles or uppers of ordinary shoes
- Molten metal that might splash on feet or legs

- Hot or wet surfaces
- Slippery surfaces

Foot and Leg Protection Choices

- Safety shoes. These have impact-resistant toes and heat-resistant soles that protect against hot work surfaces common in rooftop installation areas
- May have metal insoles to protect against puncture wounds
- May be designed to be electrically conductive for use in explosive atmospheres
- May be designed to be electrically non-conductive to protect from workplace electrical hazards

Figure 6.3: A pair of ISO 20345: 2004 compliant S3 safety boots

Figure 6.4: Non-protective footwear

6.2.5 Hand and Arm protection

Use proper hand gloves – 'Electrically Insulated Gloves' – for handling electrical connections. Two kinds of gloves are commonly used: PVC gloves and cotton gloves.

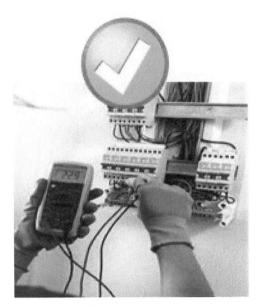

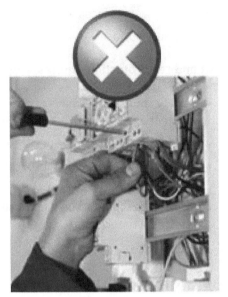

Figure 6.5: *Right Way – use protective hand gloves*

Figure 6.6: *Wrong way – Use of bare hands*

6.2.6 Safety Belt/Body Harness and Overalls (Full-Body Suit)

Safety Belt/Harness provides the following support:

- Personal protection against falling from high structures
- Enables comfortable working position with protection against slipping or imbalance
- Climbing to a location which is inaccessible from inside the building/household using an anchorage and suspension line

Overalls/protective PVC-coated jackets provide protection from:

- Extreme or harsh weather conditions
- Injury from sharp tools to the body
- Chemicals/fluids which should not come in contact with the body

Figure 6.7: Snapshot of all PPE

6.3 IDENTIFICATION AND MITIGATION OF SAFETY HAZARDS

6.3.1 Overview

Any PV system is designed to fulfill a specific load requirement. Almost all grid-connected PV arrays use hundreds of PV modules having both series and parallel connection to generate large amount of electrical power. Operating voltages may be greater than 600 Vdc and currents may be hundreds of amperes!!! Many off-grid PV systems have lesser number of modules but they use battery bank to store energy. Generally, each unit of battery is of 12 Vdc and can produce currents which can be hazardous causing severe burns.

Types of Hazards associated with PV installations

The safety considerations during an installation can be understood with respect to the various kind of hazards which could exist. These are broadly classified as in the following flowchart in Figure 6.8.

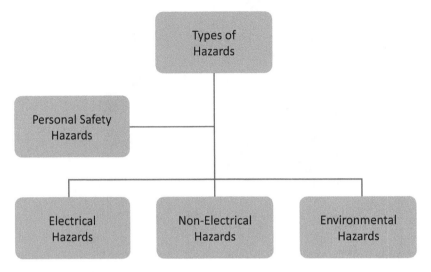

Figure 6.8: Types of Hazards associated with Solar PV installations

6.3.2 Personal Safety Hazards

This has been elaborately covered in the previous unit on 'Use and Maintain Personal Protective Equipment (PPE)'.

Non-electrical hazards

There is an improper opinion among many persons that no one can get hurt while working on a small solar PV system. But the reality is not so. One can get injured while working on any configuration of PV system. Thus, safety should be the first and foremost point to be kept in mind.

Some common non-electrical hazards are as follows:

- Exposure to sunlight because
  - PV systems are installed at the locations where the radiation is very good and no shading exits. This may cause severe sun burns (especially in the summer seasons)
- Bites of insects, snakes, and other vermin because
  - Spiders and other insects often inhabit the junction boxes
  - Snakes use the shade built by the PV array
  - Ants are also found under PV arrays or near battery storage boxes
  - Some wasps build nests in the array framing

- Cuts and bumps because
  - Most of the PV systems include metal framing, junction boxes, cables, nuts and bolts, etc
  - Many of PV installation components have sharp edges and can injure you if you are not cautious
- Falls and strains when working at height because
  - Walking with heavy loads for installation in remote areas
  - During strong winds, PV module behaves like a wind-sail and can make you fall from the ladder
- Burns due to high metal temperature because
  - Metallic surfaces which are left in sun can attain higher temperatures (~ 80 °C) which can cause burns
  - Concentrating solar PV system can also cause burns

6.3.3 Electrical Hazards

The most common electrical mishaps lead to electrical shocks or/and burns, contraction of muscles, and other severe injuries linked with falling after the electric shock. These injuries can take place anytime during the entire process of PV installation. It is difficult to estimate the severity of electrical injury because the resistivity of human body varies from 1000 ohms to several 1000 ohms. This variation in resistivity depends upon the skin moisture at the time of accident. Even a very less current (of order mA) is also sufficient to cause damage. A list of DC and AC current (in mA) and their linked-up electric shock hazards are given in Table 6.1.

Table 6.1: Common Tool Hazards

Reaction After The Electric Shock	Current	
	DC	AC
Perception: tingle, warmth	6 mA	1 mA
Shock: retain muscle control, reflex may result into injury, burns	9 mA	2 mA
Severe Shock: lose in muscle control, can't let-go, severe burns, asphyxia	90 mA	20 mA
Ventricular Fibrillation: may cause death	500 mA	100 mA
Heart Frozen: temperature of human body rises, death occurs in minutes	Greater than 1 A	Greater than 1 A

6.3.4 Battery Hazards

Any PV system with battery/ battery bank is a potential hazard. The major areas of concern are:

- Electrical burns
    - Shorting the terminals of any typical battery of a PV system may cause severe burns and even death (electric shock can occur even at low battery voltage)
- Acid burns
    - Acid of any type of battery can create burns if it get in touch with uncovered skin
    - Battery acid's contact with human eye can result in blindness
- Fire or Gas explosion
    - Most of the batteries used in solar PV installations release hydrogen gas during their charging
    - This is a flammable gas and can cause gas explosion and fire

6.3.5 Inverter Hazards

Inverter should be installed at a place where people can't reach frequently because during operation the surface temperature is very high and can cause a potential burn hazard. Ensure that the temperature of location should be in the range of -25 to +65 degree.

6.3.6 AC Power Hazards

If AC power output is required, then PV inverter is needed to be installed for the conversion of DC power from PV array/modules to AC power. PV inverter operates at high voltage both at the input and the output ends, therefore, there is enough current to cause electric hazards. Hence, use of PPE is very critical for personal safety.

6.3.7 Preventive Measures to be Taken by a PV Installer

A PV installer should always be cautious at site and follow safety measures. In this section of book, the preventive measures are detailed in accordance with the installation procedure.

Following are the general preventive measures that should be taken by a PV Installer throughout the PV system installation:

- Identify and understand the company policies required for work place safety
- Identify the right person to contact if some accident happens
- Identify the requirements of safe work
- Implement the safe work flow procedure
- Use and maintain PPE
- Know the location of First Aid box
- Read and understand the proper usage of tools/equipments
- Do proper tagging and markings of equipment
- Add labels or warning signs wherever required for the benefit of technicians as well as customer

6.3.8 Mitigation Measures to be Taken after the Occurrence of Hazards

In spite of all precautions/ preventions, there are chances of hazards. Therefore, a PV installer should always be aware of the mitigation measures to be taken if some mishap occurs. Following are the important mitigation measures that should be taken after the occurrence of hazards:

STEP 1: Rush towards the First-Aid area

STEP 2: Apply First Aid on burns/injuries/wounds

STEP 3: Apply Sodium Hydroxide (NaOH) solution when burns are due to battery's acid

STEP 4: Take the injured person to the medical room

STEP 5: Look for the supervisor/ team leader

STEP 6: Call the doctor/ medical officer

6.4 WORK HEALTH AND SAFETY AT HEIGHTS

- Perform regular maintenance
- Use right tool for the job
- Inspect all tools before use
- Use the right PPE
- Report to your supervisor of any unsafe tool

Reassess risk control measures, as required, in accordance with changed work practices and/or site conditions and undertake alterations.

- The first safety rule to keep in mind when working with PV panels or other PV components is: always stop working in bad weather.
- Never work when it's raining, immediately after rain or in wet or slippery conditions or with wet tools.
- PV panels can be blown around by the wind or a storm which can result in you falling or causing damage to the PV system.
- Do not apply pressure on PV panels by sitting or stepping on them or they might break and cause bodily injury, electrical shock, or damage to the solar panels. Also never drop anything on the PV panels.
- Make sure your entire PV system is properly and safely earth-grounded to prevent electrical shock and injury.
- Cover your PV solar panels with an opaque material during wiring to stop or prevent electricity production.
- Do not use artificial or magnified light on the PV solar panels
- Inspect all your power tools to ensure that they are working safely prior to starting the installation of your PV system
- Always get a second person to securely hold ladders as you climb and use rubber latter mats to prevent the ladder from slipping.

6.4.1 Inspect/Install Fall Protection

Fall Protection Equipment

Fall protection equipment is used in situations where a potential to fall cannot be avoided. Equipment used will both protect the fall and absorb some energy of the fall. Examples would be the traditional harness plus lanyard incorporating energy absorbance, safety nets, air bags.

Fall Protection Systems

A fall protection system should be fitted such that there is adequate clearance for it to deploy and to prevent the installer from hitting an obstruction or the ground before the fall is stopped. Many recognised practices specify either a guard rail system, safety net system, or personal fall protection system to protect the installer when exposed to a fall of 1.8 m or more from an unprotected side or edge.

Fall protection procedures should provide for a rescue to be carried out if the installer is left suspended from the working place. Safety nets or airbags should be located as close as possible to the working level to enhance their effectiveness. Safety nets should be installed as close as practicable under the walking/working surface on which surveyors are working.

A fall protection system should not be used in a manner:

i.   Which involves the risk of a line being cut

ii.   Where its safe use requires a clear zone (allowing for any pendulum effect)

iii. Which otherwise inhibits its performance or renders its use unsafe

6.4.2 Use of Safety Signs

Figure 6.9: General Safety Signs

Figure 6.10: Prohibitory safety signs

UNIT 7 ANNEXURE 1 – SAMPLE TEST CERTIFICATES OF SOLAR PV MODULES AND GRID-CONNECTED INVERTER

IEC 61215 for Solar PV Modules Design

IEC 61730 for Solar PV Modules Safety

Certificate

Registration No.: PV 60025583   Page 2   **Report No.: 15032227.002**

License Holder:
YINGLI ENERGY (CHINA) COMPANY LIMITED
No. 3399 Chaoyang North Road
Baoding 071000, Hebei
China

Product:
PV Module
Type.
YL190P-23b, YL195P-23b
YL190P-23b/2, YL195P-23b/2
YL220P-26b
YL245P-29b
YL245P-29b/2
YL270P-32b
YL290P-35b, YL295P-35b

Manufacturing Plant:
YINGLI ENERGY (CHINA) COMPANY LIMITED
No. 3399 Chaoyang North Road
Baoding 071000, Hebei
China

Basis:

☒ IEC 61730-1:2004
IEC 61730-2:2004
EN 61730-1:2007
EN 61730-2:2007
"Photovoltaic (PV) module safety
qualification"

- Qualified, IEC 61215
- Safety tested, IEC 61730
- Periodic Inspection

☒ **Factory Inspection**
To document the consistent quality of
the product factory inspections are
performed periodically.

Remarks:
- IEC EN 61730 consists of part 1 (Requirements for construction) and part 2 (Requirements for testing).
- The above listed PV modules fulfil the requirements of Application Class A (Safety Class II). They may be used in PV plants
  at a maximum system voltage (Voc at STC) of up to 1000 VDC.
- The fire test (IEC 61730-2 / MST 23) was not performed.

- Additional power ratings.
- The details of the factory inspection are documented in report no. 15031411.001.

Conditions:
The product test is voluntarily according to technical regulations. Any change of the design, materials, components or
processing may require the repetition of some of the qualification tests in order to retain type approval.
The certificate is valid until 05 May 2014.

Cologne, 09 February 2010

Dipl.-Ing. M. Adrian

TÜV Rheinland LGA Products GmbH, Tillystrasse 2, D-90431 Nürnberg / Contact: + 49 221 806 2477 email: enertest@de.tuv.com

IEC 62116:2008 for Grid Connected Inverter

SGS

Test Certificate
N° 2210 / 1230-IEC-E3 - CER

SGS Tecnos, S.A., Electrical Testing Laboratory accredited, according for testing of "SAFETY OF ELECTRICAL / ELECTRONICS, ME & IT EQUIPMENTS" and "ELECTROMAGNETIC COMPATIBILITY".
CERTIFIES that the next products:

Equipment Description..........:	SOLAR INVERTERS
Trade Mark............................:	ABB Oy
Models ....:	PVS800-57-0100kW-A(1); PVS800-57-0250kW-A(2), PVS800-57-0500kW-A(2); PVS800-57-0630kW-B(3); PVS800-57-0315kW-B(3), PVS800-57-0750kW-A(3); PVS800-57-0875kW-B(3) and PVS800-57-1000kW-C(3).
(1):Model fully tested (2):Models partiality tested (3):Extension models	
Control version ....:	ISXR7330
Ratings....................................:	**MODEL: PVS800-57-0100kW-A** Input: 450-825 Vdc (1000 Vdc MAX.) 230 A MAX. Output: 300 V 3 ~ 50/60 Hz 100 kW 195 A **MODEL: PVS800-57-0500kW-A** Input: 450-825 Vdc (1000 Vdc MAX.) 1145 A MAX. Output: 300 V 3 ~ 50/60 Hz 500 kW 965 A **MODEL: PVS800-57-0630kW-B** Input: 525-825 Vdc (1000 Vdc MAX.) 1230 A MAX. Output: 350 V 3 ~ 50/60 Hz 500 kW 1040 A **MODEL: PVS800-57-0250kW-A** Input: 450-825 Vdc (1000 Vdc MAX.) 575 A MAX. Output: 300 V 3 ~ 50/60 Hz 250 kW 485 A **MODEL: PVS800-57-0315kW-B** Input: 525-825 Vdc (1000 Vdc MAX.) 615 A MAX. Output: 350 V 3 ~ 50/60 Hz 250 kW 520 A **MODEL: PVS800-57-0750kW-A** Input: 450-825 Vdc (1100 Vdc MAX.) 1710 A MAX. Output: 300 V 3 ~ 50/60 Hz 750 kW 1445 A **MODEL: PVS800-57-0875kW-B** Input: 525-825 Vdc (1100 Vdc MAX.) 1710 A MAX. Output: 350 V 3 ~ 50/60 Hz 875 kW 1445 A **MODEL: PVS800-57-1000kW-C** Input: 600-850 Vdc (1100 Vdc MAX.) 1710 A MAX. Output: 400 V 3 ~ 50/60 Hz 1000 kW 1445 A
Applicant & Manufacturer ....:	ABB Oy Drives
Address ... .:	Hiomotie, 13. - P.O. Box 184. 00381 HELSINKI (Finland)

Were tested and found to be in **CONFORMITY** with the applicable requirements specified in the next standard:

-**IEC 62116: 2008**, Test procedure of islanding prevention measures for utility-interconnected photovoltaic inverters.

This certificate, issued by SGS Tecnos, is laid down upon results, analysis y verifications detailed in the reference Test Report n° 2210/1230-IEC-E2.

Madrid, 22th March, 2013

SGS Tecnos, S.A.
Laboratorio de Ensayos E&E
FERNANDO MONTES CLAVER
Technical Manager of the E&E Lab

IEC 60068-2-6 for Grid Connected Inverter

ABB

Statement of Compliance

Manufacturer: ABB Oy

Address: P.O Box 184, FIN-00381 Helsinki, Finland.
                           Street address: Hiomotie 13

herewith state under our sole responsibility that the solar inverter series for photovoltaic
power systems with type marking:

                           PVS800-57

to which this statement relates, is in conformity with applicable requirements of IEC 60068-2
(Envirionmental testing). More detailed information is available in relevant product
documentation.

Following standards have been referenced:

                           IEC 60068-2-6: Environmental testing – Part 2-6:
                           Tests – Test Fc: Vibration (sinusoidal)

                           IEC 60068-2-29: Basic environmental testing
                           procedures –
                           Part 2-29: Tests – Test Eb and guidance: Bump

Helsinki, 2011-07-07

Ilkka T Ikonen
Vice President
ABB Oy, Drives

IEC 61683 for Grid Connected Inverter

Intertek

Reference No: 100633J-ELSA   1(4)

Statement of test results

for efficiency measurements

Product:	PVS800-57, Photovoltaic central inverter
Tested on request of:	ABB Oy Drives, P.O. Box 184, FI-00381 HELSINKI, FINLAND
Name and address of the factory:	ABB Oy Drives, Hiomotie 13, FI-00380 HELSINKI, FINLAND
Rating and principal characteristics:	Input: Umppt(DC) 600…850 V, Umax(DC) 1100 V, Imax(DC) 1710 A at +50°C
	Output: PN(AC) 1000 kW, UN(AC) 400V, IN(AC) 1445 A at +50°C, Imax(AC) 1732 A at +25°C, f 50/60 Hz
Trade mark:	ABB
Model/Type reference:	PVS800-57-1000KW-C
Test results:	See pages 2-4.
Additional information:	The test setup has been in accordance with standard IEC 61683:1999 (1st Edition) and with EN 50530 with test results reported in test report reference no. 100633C-ELSA. Efficiency measurements and calculations done up to Umppt(DC) 875 V

This Statement is only valid for the tested sample.

Hyvinkää January 23. 2013

Mika Virtanen
Testing Engineer

Intertek ETL Semko Oy
Koneenkatu 12 (Door K1)
FI-05830 Hyvinkää Finland
Business ID: 1871439-4
Phone +358 10 424 6200
Telefax +358 10 424 6201

UNIT 8 ANNEXURE 2 – GUIDELINES FOR GRID SOLAR ROOFTOP PV SYSTEM ON NET-METERING BASIS FOR BESCOM OFFICIALS

8.1 INTRODUCTION

This document describes the general conditions and technical requirements for connecting Solar Rooftop Photovoltaic (SRTPV) system installations to BESCOM grid in accordance with the provisions provided by Indian Electricity Act 2003, Karnataka Solar policy 2014-21 dated: 22.05.2014 and KERC's tariff order dated 10.10.2013 and distribution code approved by KERC.

8.2 PROCEDURE FOR INSTALLATION AND COMMISSIONING OF THE SRTPV SYSTEM

a. Registration of Application

b. Processing of Application

c. Approval letter for submission of technical details of proposed SRTPV system (Pre-installation)

   (To be intimated within 7 working days from the date of registration)

d. Approval for installation

   (To be intimated within 3 working days from the date of submission of technical details)

e. General guidelines

f. Signing of power purchase agreement

g. Submission of work completion report along with necessary approvals (Approvals from Chief Electrical Inspectorate, GoK, are to be given within 7 working days from the date of submission)

h. Communication of approval for testing, commissioning, and synchronising (On receipt of documents, approval shall be given as per Format-7 within 3 working days from the date of receipt of documents)

i. Commissioning and synchronising of SRTPV system installation

(To be processed within 3 working days from the date of approval)

j. Billing procedure

k. Periodical inspections

a. Registration of application:

1. The application can be downloaded from BESCOM website (www.bescom.org) (Format-1).

2. The address and details of sub-division and its code can be downloaded from the BESCOM website.

3. Applications can also be registered online. Copy of the online application registered along with application fee and necessary document shall be submitted to the concerned C,O&M subdivision, AEE, BESCOM.

4. The filled-in application along with necessary documents has to be submitted to concerned C,O&M sub-division, BESCOM. The registration fee payable are as follows:

Sl. No.	Sanctioned Load/Contract Demand	Registration Fee
1.	Upto and inclusive of 5kWp	₹500/-
2.	Above 5kWp and below 50kWp (67Hp/59kVA)	₹1000/-
3.	Above 50kWp (67Hp/ 59kVA) and upto 500kWp.	₹2000/-

5. Assign the application registration number with acknowledgment.

6. Separate application register has to be maintained at sub-divisional office for solar rooftop installations (Format-2).

b. Application Processing:

1. After registration, application will be forwarded to revenue section for verification of name, RR No., sanctioned load, tariff, arrears if any, etc. Verification shall be completed within 2 working days (Format-3).

2. After verification by the revenue section, the application will be sent to concerned section officer for spot inspection and submission of technical feasibility report. The report shall be submitted within 3 working days from the date of receipt of SRTPV application (Format – 4).

3. If installation is technically feasible, letter addressed to the applicant is sent to submit the technical information of all the equipments proposed to be used for SRTPV systems (Format – 5).

4. Checks before issuing the technical feasibility of proposed interconnection:

   i. The transformer shall be loaded (proposed SRTPV system load) upto 80% of capacity of distribution transformer.

   ii. To check whether the proposed interconnection will require upgrading the capacity of existing distribution network.

   iii. Phase balancing has to be checked to avoid unbalancing of load in the secondary circuit of distribution line.

c. Approval letter for submission of technical details of proposed SRTPV system to be issued within 7 working days from the date of registration of application.

d. Approval for installation:

1. All the technical information like Make, Type, etc., of the SRTPV equipments are to be checked and the test reports and certificates are to be verified in detail.

2. If technical information furnished comply with the condition stated in Format – 5, approval letter for installation of SRTPV system will be issued by the competent authority (AEE/EE) within 3 working days from the date of submission of technical report (Format – 6).

e. General guidelines:

1. The applicant is required to install the SRTPV system through system installer who have experience in design, supply, and installation of SRTPV system.

2. The agency shall assist in obtaining approval from Chief Electrical Inspectorate, Government of Karnataka, to meet safety standards and to procure net meter as per CEA guidelines from BESCOM approved vendors.

3. Inverters of MNRE approved manufacturers shall be used. The list of MNRE approved inverter manufacturers is available at the BESCOM/ MNRE website. Only those inverters which meet all required IS/IEC standards shall be eligible for installation.

4. Data monitoring: Online monitoring will be compulsory for all the systems of more than 50kWp capacity. The SRTPV plant parameters are measured and transmitted to ALDC, BESCOM, using SCADA system.

5. The Applicant/System installer shall obtain approval of drawing from Chief Electrical Inspectorate (CEI) of Government of Karnataka (GoK) before commencing installation work for installations above 10kWp.

6. The SRTPV system should comply with the relevant (BIS/IEC) technical standards.

7. The installation work has to be carried out as per the approved drawing and standards.

8. In case the installed (also read proposed) capacity of the SRTPV system is higher than the sanctioned load of the consumer, which consequently requires an upgradation in the infrastructure (service line, meter with CT, if required), transformer upgrading (if required), the consumer will have to upgrade at his/ her/at her own cost.

9. Work completion report along with required documents to be submitted within 180 days from the date of issue of approval letter for installation to the concerned AEE, C,O&M sub-division, BESCOM along with receipts of facilitation fee as follows:

Sl. No.	Sanctioned Load/Contract Demand	Facilitation Fee
1.	Upto and inclusive of 5kWp	₹1000/-
2.	Above 5kWp and below 50kWp (67Hp/59kVA)	₹2000/-
3.	Above 50kWp (67Hp/ 59kVA) and upto 500kWp	₹5000/-

10. The existing metering wiring shall be changed to solar power generation side in presence of AEE/EE, MRT, BESCOM to measure solar generation.

11. The applicant has to procure bi-directional meter from any of the approved vendors of BESCOM and the meter has to be tested by MT division, BESCOM and the same shall be fixed at interconnection point.

12. The applicant shall provide check meters when the SRTPV system is more than 20kWp.

f. Signing of power purchase agreement:

1. After completion of SRTPV installation work, the consumer has to enter into a power purchase agreement with BESCOM on ₹200/- stamped paper.

2. The AEE, C,O&M sub-division is the signing authority for PPA upto sanctioning load of 49kWp and EE, C, O&M division is the signing authority for PPA of sanctioning load of above 50kWp.

3. Copy of the PP agreement shall be submitted to the General Manager (Power Purchase), Corporate Office, BESCOM, Bangalore.

g. Submission of work completion report:

The applicant/system installer of SRTPV system shall submit the following documents along with work completion report as per Format – 6A to the approving authority (C,O&M, AEE/EE of BESCOM):

a. Approved drawing and approval letter for commissioning the SRTPV system by CEI of GoK.

b. Specification sheets of all equipments and manufacturer's test reports and test certificate of modules and inverters.

   c.　Test certificates of bi-directional meter from MT division, BESCOM.

   d.　Undertaking for obtaining MNRE subsidy from KREDL or self-declaration for not obtaining the MNRE subsidy (Format-1C).

   e.　Details of facilitation fee paid.

   f.　Power purchase agreement on Rs. 200/- stamp paper.

h.　Communication of approval for commissioning and synchronising:

   1.　All the documents (a–g) in the above para are to be verified in detail by the competent authority (C,O&M, AEE/EE).

   2.　After verification of above documents, the sanctioning authority has to issue approval letter for commissioning and synchronising the SRTPV system with BESCOM grid within 3 working days from date of receipt of all documents (Format-7).

i.　Commissioning and synchronising of SRTPV system installation:

   1.　C,O&M, AEE/EE has to inspect the PV modules connections, earthing, isolating switches, functions of inverter, sealing of the energy meters, meter boxes, recording of readings, preparation of testing, and commissioning reports.

   2.　Earthing protection: Both equipment earth (DC) and system earth (AC) to be checked for proper earthing.

     •　System earth: means used to ground one leg of the circuit.

      For ex: in AC circuits the neutral is earthed, while in DC supply, +ve is earthed.

     •　Equipment earth: All the non-current carrying metal parts are bonded together and connected to earth to prevent shocks to the manpower and protection of the equipment.

     •　Lightening arrester

   3.　Surge protection:

     •　Surge protection shall be provided on the DC and AC side of the solar system.

     •　The DC (SPDs) shall be installed in the DC distribution box adjacent to the solar grid inverter.

     •　The AC SPDs shall be installed in the AC distribution box adjacent to the solar grid inverter.

     •　The SPDs earthing terminal shall be connected to earth through the above mentioned dedicated earthing system. The SPDs shall be of Type 2 as per IEC 60364-5-53.

4. The PV module structure components shall be electrically interconnected and shall be grounded.

5. Earthing shall be done in accordance IS 3043-1986, provided that earthing conductors shall have a minimum size of 6.0 mm2 copper, 10 mm2 aluminum, or 70 mm2 hot dip galvanised steel. Unprotect aluminum or copper-clad aluminum conductors shall not be used for final underground connections to earth electrodes.

6. A minimum of two separate dedicated and interconnected earth electrodes must be used for the earthing of the solar PV system support structure with a total earth resistance not exceeding 5 ohm.

7. The earth electrodes shall have a pre-cast concrete enclosure with a removal lid for inspection and maintenance. The entire earthing system shall comprise non-corrosive components.

8. The synchronisation of the SRTPV system shall be carried out by the concerned sub-divisional/divisional engineer along with MT staff within 3 working days from the issue of approval for synchronising and commissioning.

j. Periodical inspections:

1. The meters both uni-directional and bi-directional are to be tested as per Schedule by MT staff once in 6 months as per KERC norms.

2. The inverter functionality of every installation is to be checked by MT staff of BESCOM once in 6 months.

3. Periodical test reports/inspection reports shall be submitted to the concerned C,O&M sub-divisional office.

k. Billing procedure:

1. The consumer shall receive a monthly net import/export bill indicating either net export to the grid or net import from the grid.

2. 'Import'- means energy supplied by the BESCOM grid.

3. 'Export'- means energy delivered to the BESCOM grid.

4. The meter reader has to capture present reading of uni-directional meter provided at solar side.

5. Solar side meter reading is used only for MIS report (to measure the quantum of generation) and shall not be used for billing purpose.

6. The meter reader has to capture import & export energy and other billing parameters recorded by the bi-directional meter.

7. In case of net import bill, the consumer shall pay the same as per existing tariff.

8. In case, the export energy is more than the import, AEE, C,O&M sub- division will arrange for the energy exported as per the KERC approved tariff through NEFT.
   - Without subsidy Rs.9.56
   - With MNRE subsidy Rs.7.20

9. Minimum charges/electricity dues/statutory levies, if any, shall be adjusted against the energy purchase bill.

10. The amount payable by the BESCOM to the seller for energy injected to the BESCOM grid (excluding self-consumption) during the billing period becomes due for payment, which shall be settled within 30 days from the date of meter reading and will be credited to the bank account through NEFT.

UNIT 9 ANNEXURE 3 – GUIDELINES FOR UTILITY PERSONNEL JVVNL DOCUMENT FOR OWN EMPLOYEES

9.1 INSTALLATION PROCESS OF SRTPV SYSTEMS

Procedure for installation and commissioning of the SRTPV system:

a. Registration of application form

b. Approval for installation

c. Submission of documents (Post- installation)

d. Approval for testing, commissioning, and synchronising

e. Commissioning and synchronising of SRTPV system installation

f. Periodical inspections

g. Billing procedure

a. Registration of Application

1. The application can be downloaded from JVVNL website (http://www.jaipurdiscom.com/) OR it can also be obtained from the sub-divisional offices of JVVNL.

2. The filled-in application along with necessary documents, application fees, and security deposit has to be submitted to concerned department of JVVNL. The application fee payable to different consumer categories are as follows:

S. No.	Description	Amount
1.	Application Fee	
	i. LT Single Phase	₹200
	ii. LT Three Phase	₹500
	iii. HT – 11 kV	₹1000
	iv. HT – 33 kV	₹2000

S. No.	Description	Amount
2.	Security Deposit for Solar PV Plant	
	(a) Domestic	₹100/kW
	(b) Non-domestic and others	₹200/kW

3. The concerned sub-divisional office shall assign the application registration number with acknowledgment. Separate application register has to be maintained at the sub-divisional office.

b. Submission of Documents (Pre-Installation)

While submitting the application form the consumer also needs to furnish additional information in the form of attached documents. The documents required to be submitted with application forms are as follows:

1. Electricity bill copy
2. Authorisation letter, if applicant is other than an individual
3. Subsidy application / sanction letter OR Self-certification of not availing subsidy

Within ten (10) working days of receipt of the eligible consumer's application, the distribution licensee shall provide written notice (via email or letter) that it has received all documents required by the standard interconnection agreement or indicate how the application is deficient.

c. Application Screening

After receiving the application, the concerned department of utility will examine the application on the basis of data sufficiency and technical feasibility of interconnection as per set technical standards and parameters; the process followed by the utility for screening of application is as follows:

1. Utility authorised personnel will review the contact details and other relevant information sought in the application format such as applicant name, address, project capacity, consumer category, banking details under the application form.
2. On completion of general screening, utility will check whether the capacity of proposed SRTPV facility is feasible with respect to the available capacity of service line, distribution line, distribution transformer, and protective devices.

3. If required, utility will go for interconnection study to check the feasibility of proposed interconnection. In this case, utility will intimate feasibility or otherwise within sixty (60) days from the receipt of completed application.

4. On accessing the feasibility, utility shall intimate the eligible consumer within fifteen (15) days from the receipt of completed application. The feasibility shall be valid for a period of one month, unless extended by the distribution licensee.

5. While intimating the feasibility for the connection of Rooftop PV Solar Power Plant, the distribution licensee shall also intimate the eligible consumer:

   a. The details of documents to be submitted by the eligible consumer.

   b. Particulars of any deficiencies, if noticed, along with instructions to remove such defects.

   c. Details of any interconnection study required.

   d. The amount of security deposit for the installation of the Rooftop PV Solar Power Plant.

6. After successful completion of technical review, utility will give go ahead certificate to consumer and will ask to furnish technical information in the form of project completion report.

d. Approval for Installation

1. After registration, application will be forwarded to the revenue section for verification of name, signature, sanctioned load, arrears if any, etc.

2. Then the application will be sent to the concern officer for spot inspection and submission of technical feasibility report within 3 days from the date of receipt of SRTPV application.

3. If the interconnection application passes all the criteria of utility screening process, then utility will approve the interconnection request and will send a written notification to applicant within 10 working days from the date of receipt either via email or post.

4. If the application form fails to clear the general or technical screening, then utility will send notice/intimate the applicant, the reason for failure and to provide additional details (if required) to process the application form.

   a. Applicant within seven (7) working days of receiving the notice from the utility shall respond with all the required information to the utility.

   b. If the applicant is unable to provide the required information to the utility within stipulated time period, then applicant may request utility for extension of time by writing a request letter to utility.

    c. If the applicant neither responds to utilities notice nor request for extension of time lines within 30 days of receiving the notice, then the application will be deemed withdrawn.

5. If the applicant is required to install the SRTPV system through agencies who have experience in design, supply, and installation of SRTPV system, the agency shall assist in obtaining approval from Chief Electrical Inspectorate, Government of Rajasthan, to meet safety standards and to procure net meter as per CEA guidelines from JVVNL/RRECL, Govt. of Rajasthan approved vendors.

6. The installation agency has to obtain approval of drawing from Chief Electrical Inspectorate (CEI) of Government of Rajasthan (GoR) before commencing installation work.

7. The SRTPV system should comply the relevant technical standards, as specified under the RERC net-metering regulations.

8. The installation work has to be carried as per the approved drawing and standards.

9. Work completion report along with required documents (post approval) to be submitted within 60 days from the date of issue of approval letter for installation to the concerned department of JVVNL along with receipts of facilitation fee.

10. The existing metering system shall be shifted to generation side of SRTPV to measure solar power generation.

11. The consumer has to procure bi-directional meter from any of the approved vendors of JVVNL and the meter has to be tested by MT division, JVVNL, and the same is to be fixed at interconnection point.

12. The applicant shall complete the installation of rooftop facility within 60 working day from getting the approval from the utility and shall have to intimate the distribution utility for conducting commissioning test before interconnection.

    a. In case the applicant fails to complete the installation of rooftop facility within 60 days, it may request for extension of time period citing reasons for delay.

    b. If the applicant neither seeks extension of timelines nor intimate the utility for commissioning test within 60 days, then it will be considered as withdrawn by the applicant.

    13. On getting intimation for commissioning test from the applicant, utility will have to respond to it within 7 working days and it will have to communicate the proposed date for conducting commissioning test to applicant.

e. Submission of Documents (Post-Installation)

After completion of SRTPV installation work, the consumer has to submit the following documents to the approving authority (AEn/EE of JVVNL):

1. Approved drawing from CEI of GoR
2. Test certificates and reports of PV modules, inverters, cables, etc.
3. Test certificates of bi-directional meter from MT division, JVVNL
4. Work completion report
5. To pay the prescribed facilitation fees
6. Power purchase agreement

f. Approval of Testing, Commissioning, and Synchronisation

1. All the documents submitted are to be verified in detail by the competent authority.
2. To visit the premises and to inspect the SRTPV equipment, circuitry, meters, earthing, etc. along with the inspection team of the Chief Electrical Inspectorate, GoR.
3. The Chief Electrical Inspectorate, GoR, will provide completion letter for synchronizing and connecting the SRTPV system with JVVNL grid.
4. The sanctioning authority has to issue approval letter for commissioning and synchronising the SRTPV system with JVVNL grid on production of approval letter from CEI, GoR.

g. Commissioning and Synchronisation of SRTPV System Installation

1. JVVNL has to inspect the PV modules connections, earthing, isolating switches, functions of inverter, sealing of the energy meters, meter boxes, preparation of testing, and commissioning reports.
2. JVVNL has to complete the security measure checklist before performing the test run and will share the copy of it to the applicant for future reference.
3. Earthing protection: Both system earth and equipment earth to be checked for proper earthing.
   a. System earth is used to ground one leg of the circuit. For example: in AC circuits the neutral is earthed, while in DC supply +ve is earthed.

b. Equipment earth: All the non-current carrying metal parts are bonded together and connected to earth to prevent shocks to the manpower and protection of the equipment.

4. JVVNL staff has to conduct test run before actual commissioning of the system.

5. The synchronisation of the SRTPV system shall be carried out by the concerned Sub-divisional/Divisional Engineer along with MT staff.

h. Periodic Information

1. The meters are to be tested as per schedule by MT staff.

2. The inverter functionality of every installation is to be checked by MT staff of JVVNL once in 6 months.

3. Test/Inspection reports to be submitted to the concerned sub-divisional office.

i. Billing Procedure

1. The consumer shall receive a monthly net import/export bill indicating either net export to the grid or net import from the grid.

2. 'Import'- means energy supplied by the JVVNL grid.

3. 'Export'- means energy delivered to the JVVNL grid.

4. The meter reader has to capture present reading of uni-directional meter provided at solar side.

5. Solar side meter reading is used only for MIS report (to measure the quantum of generation) and shall not be used for billing purpose.

6. The meter reader has to capture import & export energy and other billing parameters recorded by the bi-directional meter.

7. In case of net import bill, the consumer shall settle the same as per existing tariff.

8. In case, the export energy is more than the import, JVVNL shall pay for the energy exported as per the RERC approved solar tariff.

9. Minimum charges/Electricity dues/Statutory levies, if any, shall be adjusted against the energy purchase bill.

10. The net payable (credit balance if any) shall be carried forward from month on month till the accumulated energy credits are less than 50. On accumulation of 50 energy credits, JVVNL shall settle the account by paying for 50 energy credits as per RERC net-metering regulations.

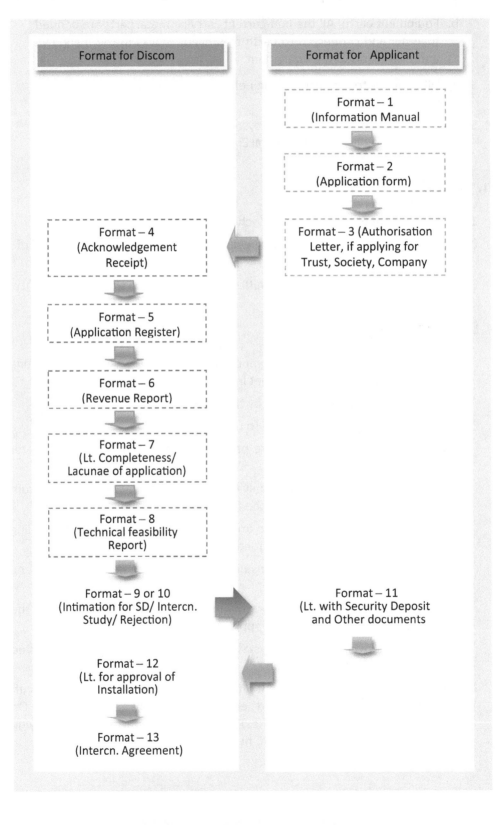

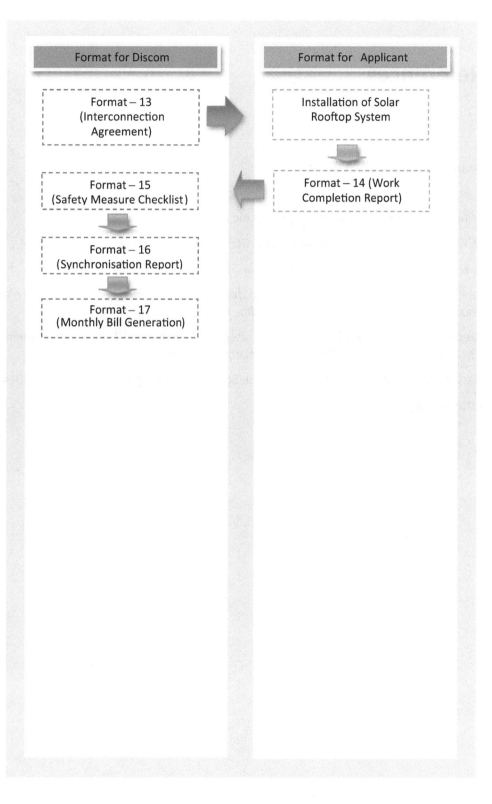

References

Biomass Energy Basics, www.nrel.gov

Solar Radiation Resource Terms, www.nrel.gov

Ministry of Power, GOI, www.powermin.nic.in

Ministry of New and Renewable Energy, GOI, www.mnre.gov.in

Sawin, J.L. et al., 'Renewables 2016 Global Status Report', REN21, 2016

Jani, O. et al., *Best Practices Manual for Implementation of State-Level Solar Photovoltaic Rooftop Programmes in India*, 1st Ed., 2016

Teece J.D., 'Business Models, Business Strategy and Innovation', *Long Range Planning* 43 (2010) 172e194; http://www.elsevier.com/locate/lrp

Garg A. et al, 'Harnessing Energy from the Sun: Empowering Rooftop Owners', 2014

https://npp.gov.in